Creative Experimenting Using Rubber Bands

For adventurous minds of all ages!

Version 2.4

By
David Tracy

authorHOUSE™

1663 LIBERTY DRIVE, SUITE 200
BLOOMINGTON, INDIANA 47403
(800) 839-8640
WWW.AUTHORHOUSE.COM

AuthorHouse™ 1663 Liberty Drive, Suite 200 Bloomington, IN 47403 www.authorhouse.com Phone: 1-800-839-8640	AuthorHouse™ UK Ltd. 500 Avebury Boulevard Central Milton Keynes, MK9 2BE www.authorhouse.co.uk Phone: 08001974150

First published by AuthorHouse 4/03/07

ISBN: 1-4259-2754-8 (sc)

Printed in the United States of America
Bloomington, Indiana

This book is printed on acid-free paper.

Contents:

Introduction: Science History
"What's the matter with matter?"

For some time I have been preparing a set of lessons, on various topics from material science, chemistry, and physics, to geology, astronomy, and even economics, using a common rubber band! The most important things you will need to know to perform the "creative experiments contained within the book are

> 1. A rubber band can stretch and regain its shape
> 2. A rubber band can oscillate
> 3. A rubber band can break

All of the experiments make use of these simple facts, however, there are many other ideas to consider in studying the rubber band including: chemistry, thermodynamics, electrical conductivity, and other topics which have a close relationship to it.

Then there's the question of motion–what is motion? How can I measure it?

In this text, I will draw heavily from the work of three major scientists: Galileo, Newton, and Einstein, since their works lay the foundations of what we know about matter and the universe, and how we know it! Also, a few others like the "pre-Socratic philosophers in ancient Greece, who were the first to conjecture about the ultimate nature of the material world, and the works of Gauss and Riemann, who established the beachhead for Einstein's work in relativity.

Along the way, other ideas in science will get introduced such as the theories of catastrophe and chaos, in the hope that their addition will help the reader to see the large scale motions of the universe in the micro- and macroscopic world of the rubber band.

My purpose in writing this book is to provide a way for the intellectually curious to find out for themselves how things really work by experimenting, observing, and examining the same test article (a rubber band) from many different points of view. Can a simple rubber band be like a "Eudoxan Sphere" helping us to model the *inner world* hidden inside the rubber band?

The beauty of using the rubber band for this purpose is that its own characteristics are so favorable. It has a comfortable physical size, weight, and density to be measured easily using lab scales, rulers and micrometers, while many of its important physical characteristics like its "natural frequencies" can be easily measured using a "stop watch" instead of more complex and costly equipment.

A rubber band can be used as a "*highly representative* example of matter," demonstrating all the important physical qualities, which if mediated upon by observation, can reveal just what is so remarkable about it! Rubber bands are ideal for this, there a little like finding the "primordial substance" that Thales sought in his ideas about water.

In the same way that the practical needs of builders led them to draw real lines on the curved earth for the purposes of construction, city planning, and mapping, and this practical need led mathematicians and cartographers to create "non-Euclidean geometry" (a geometry which is not based on the assumption that all parallel lines diverge at infinity), so too does the observing of the motions of rubber bands lead us to an understanding of the qualities and characteristics of matter in motion, which helps us understand how to bridge a gap from the old physics of mechanics with the new modern physics! There is an entire universe down inside a rubber band containing gas, plasma, atoms, electrons, photons and particles. These constituents can be examined closely and even allowed, by analogy, to take the character of galaxies in clusters expanding into an endless void, to model economic processes, or even earthquakes!

Achieving such new insights first demands that we shed some of our "old thinking" and make room for embracing newer, freer, and more fluid concepts in how we picture matter, but with a little practice and a lot of imagination, I'm sure the reader will get the idea!

Only then can this new understanding liberate us from the dark shadows and "*dead dogma*" of the traditional curriculum! My hope is that today's students will not be like those people of Galileo's time, who were too intimidated by superstition, to look into the telescope for fear of being possessed by evil spirits! Instead, I hope the reader will go ahead and "take a good look" at what's here, because the thing you must respect about experiment is that

"*This is what really happens!* "

Pre-Socratic philosophy: Searching for the "primordial stuff"

Who were the pre-Socratic philosophers you ask? They were the first of the early Greeks to take a new direction in their explanations of nature replacing the common one of invoking "the Gods" with a new rational model using critical thinking and observation to explain the sources of natural phenomena. This new form of discourse established the "Ionian School of thought" and became the source of the new field of "Philosophy" and marking a different form of thinking about their world from the models of the earlier "Olympian Religion." Their founding father was "Thales of Miletus" a sailor, who was born in Miletus, a small Greek city in Asia Minor (in what today is Western Turkey), who is known to have traveled extensively throughout the Greek, Egyptian, and Babylonian worlds collecting the learning of these cultures (especially their astronomy and mathematics), which led him to be the first to begin critically questioning the primacy of the old gods.

In this paper I will address the questions "What were their beliefs and major intellectual achievements?" "Do they deserve to be called the "First Scientists?" Or, do legends of their travels to Egypt and Babylon suggest that they were strongly influenced by other cultures? Their travels and very fragmentary history, leave many unanswered questions and lingering doubts about their validity, and of their originality.

But in spite of these criticisms, the philosophers of the "Ionian school" helped to really change the course of intellectual history. To explain how this happened let me begin by surveying each of these men in approximately chronological order, beginning with Thales, and highlighting their principle beliefs, contributions and accomplishments.

(The following is an excerpt from a paper entitled "Searching for the first scientists- reflections on the pre-Socratic philosophers" by the author David Tracy, which was prepared for a course in science history at Cal-State L.A.)

Thales of Miletus (f. 585 BC):

Thales is most famous for predicting a solar eclipse in the year 546 BC. As the story goes the eclipse happened during a battle between the Medes and the Lydians forcing them to lay down their arms, but there are many reasons for doubting this and suggesting that he really did nothing more than just pin the down the "year" in which this eclipse would happen. We know this because he had traveled to Babylon and probably learned how to predict eclipses from information such as "astronomical tables" that were available there. It is known that the Babylonians were interested in eclipses for astrological and religious reasons, and had been making solar observations since about 721 BC. Their tables are the only way he could've even begun to make such a prediction (Kirk & Raven, p.79-80) and this creates a controversy. "Was Thales of Miletus the founder of western science, or merely a relayer of theories developed in more distant near eastern cultures?" (McKirahan, p.25).

Another story about Thales comes from Plato:

"Once while doing astronomy, Thales was looking up towards the sky, when he fell into a well. A slave girl came along and made fun of him saying that he was "eager to know about the heavens but failed to see what was in front of him!"

This story is usually told to suggest that Thales was an "absent minded professor," even though the real Thales was distinctly pragmatic (McKirahan, p.23) (Kirk & Raven, p.79).

Yet another story about Thales tells of how to prove the value of his philosophy and to show his critics that it was not useless, he used his knowledge of astronomy and learned that there was to be a particularly large olive crop. So while it was still winter, he obtained the funds to invest in the rental of every olive press in Miletus and the neighboring city of Chios. Since no one bid against him, when the demand increased, he was thus able to rent all of the olive presses at whatever terms he liked (since he had essentially cornered the market!). He ended up making a lot of money (he made a killing!), thus showing that his philosophy was definitely not useless! And though philosophers can become easily wealthy, this is not their interest (Mc Kirahan, p.23-24).

Thales is credited with introducing geometry from Egypt. He saved their early teachings for posterity and also made a few discoveries. He invented the notion of "proof" and his two most famous theorems are:

1) The "angle-side-angle congruence theorem" (distance to a ship at sea) and 2) "similar triangles congruence" (height of a pyramid), but his most important beliefs have to do with his astronomy.

Thales cosmological views come down to just three ideas:
1) The earth floats on water, which is in some way the source of all things
2) Water is the original material.
3) All things are full of Gods.

Thales views on water as the "*original primordial material*" probably echo similar views expressed in Egyptian cosmologies, where the earth is born out of water. Aristotle later criticized Thales ideas because as he put it "There's nothing to support the water!" (Kirk & Raven, p.88)
As a final note on Thales, he is also credited with the discovery of "static electricity" and "magnetism."
He used two different types of rocks: Amber and Loadstone, which when rubbed would "Attract" certain objects to them. Thales believed that this attraction implied that the stones had "Souls" and this shows that his conversion to an "impersonal idiom" for explaining phenomena was not complete. Some describe him as a "transitional thinker" because of this lingering attempt to show that things are "full of gods" (Kirk & Kaver, p.95).

Anaximander of Miletus (f. 565 BC):

He was there to witness the "solar eclipse" predicted by his teacher; Thales in 546BC.
Anaximander is most revered not for the views he held, but for *criticizing those of his teacher*. It was this new "*Critical Attitude*" which most set the members of the Ionian school apart from the rest of the thinkers of their day.
He studied astronomy and is credited with bringing many Babylonian ideas to the Greek world including: the Gnomon (which is used for measuring the sun's shadow and can determine highest & lowest suns, as well as solstices) and the Sundial (He set one up at Sparta for indicating solstices and equinoxes. He also made "hour-makers" (the Babylonian: "twelve parts of the day")) (McKirahan, p.32).
He is also credited with making the first "Map" of the known world, even if crude and eventually improved upon by a fellow Milesian: <u>Hecateus</u> (McKirahan, p.33).
As for his physical theories, he viewed a thing called an "Arche" as the starting point of basic source and a substance called "*Apeiron*" as being the stuff out of which all things are composed. This was not just any building material but was intended as a replacement for Thales "water." Anaximander viewed the Apeiron differently than water, fire, or other proposed basic materials out of which all things come. For Anaximander the Apeiron is "undefined," but gives the suggestion that it is "eternal," "ageless," and "always in motion." That the plurality of worlds arise from it and are surrounded by it (McKirahan, p.33-34).
"Apeiron" is a compound word where the prefix- "A" means "Not" and the noun "peirar" means "Limit," so that this whole word means something like "Unlimited," or "Boundless." The word root "Per" means "Unable to get through," or "What can not be traversed end to end." So the Apeiron is 1) Infinitely large, 2) Ageless, and 3) An indefinite kind of material (since a definite material would have destroyed things with opposite characteristics!). The idea of the "Apeiron" as an indefinite substance is intended as a refutation of Thales "water," but being infinite, ageless, and surrounding everything, so that it performs the same way (McKirahan, p.33-36).
In his "Cosmological views," Anaximander taught that
1) The Stars travel on circles of fire, which have "vents" so the light may be seen though they are surrounded by mist.
2) The Sun is equal to the Earth and its "vented circle" is twenty seven times the size of the earth.
3) The Earth is cylindrical and its depth is one-third its breadth. We walk on its upper surface
4) The Moons circle is eighteen times the size of Earth.
5) The Earth is at the center of the universe and its "Obliquity" is caused by wind.

This Earth centered view of the universe would remain for many centuries of thinkers, all the way down to Copernicus and this was another significant refutation of Thales (McKirahan, p.38-43).

Anaximenes of Miletus (f. 540 BC):

Anaximander's student, he refuted his teacher by proclaiming that the underlying nature of "Apeiron" isn't indefinite, but instead Apeiron is made of "Air" differing in rarity and density according to the substances it becomes. "Finer when it becomes fire, coarser when it becomes wind, then Earth, then Stones. Everything else comes from these." (McKirahan, p.48-55) This was an objection to Anaximander!

"Air" – takes on many forms and appearances, thus accounting for the diversity of different substances from just one substance. The Greek word "Aer" means "Mist" and "Aither" then becomes the word for the bright clear part of the atmosphere, as distinct from the lower part (Mc Kirahan, p.48-55).

Anaximenes' cosmology may summed up as follows:

1) When the Air is feted the Earth is the thing to come into being.
2) The Earth is "flat" and rides on "Air." Likewise, the sun, moon, and other heavenly bodies are "flat" and carried upon "Air."
3) The stars are condensed from the moisture rising up from the Earth.
4) The constant movement of "Air" accounts for the movements of celestial bodies.
5) The Earth is supported from underneath by a column of Air.

Anaximenes, like Anaximander before him, is revered not so much for his ideas but for his refutation of his former teacher's ideas and this continuing "critical attitude" found among the thinkers of the Ionian school. The idea of a flat earth remains a popular one even today with some skeptics.

Xenophanes of Colophon (f. 570 BC):

He was a poet with wide ranging interests, born in Ionia not far from Miletus and was acquainted with its philosophers. He witnessed the Persian conquest of 546 BC and was, like most Milesians of his time, unanimous about their belief in the divine nature of the underlying stuff, but accounted for the world through natural processes and this had an unexpected consequence- it undermined the traditional religious belief in the Olympian gods. It was felt that these explanations of nature left no room for anthropomorphic Olympian gods to govern and control the natural phenomena and human destinies.

Xenophanes left no Cosmological theories, rather it is his rejection of traditional religion and his reflections on the subject of human knowledge for which he is remembered and for this he is regarded as the "Father of Epistemology" (McKirahan, p.59-68).

Heraclitus of Ephesus (b. 540 BC):

The member of an Aristocratic family, he lived to be sixty years old. His family association made him an entitled nobleman with an honorary "Kingship." Only fragments of his writing survive. An "Ionian Monist" he is credited with believing the following:

1) Fire is the basic substance.
2) The world is in a constant state of flux, change is constant and there is no "stability."
3) The universe is governed by "Logos" a word which gets many different translations like "god" or "true word," but I think it's best to think of Logos as being the "*Source of the basic fiery substance, out of which he constructed all things*" (Popper, p.147)

In the cosmology of Heraclitus the Kosmos (universe) is eternal, its material constituents are: fire, water, and earth, which systematically and regularly change into one another (McKirahan, p.128-134, 139-144)

It is the regularity of this change, which guarantees the stability of the survival of the Kosmos.

Pythagoras of Samos (Pythagorean School) (b. 570 BC):

Though he was born on the island of Samos, he left due to a disagreement with the policies of the tyrant Polycrates. It was during this time that he visited both Egypt and Babylonia, settling in Croton, a Greek city in Southern Italy where political life was dominated by "clubs" and "associations" (a little like the mafia!)
(McKirahan, p.79).

The "Pythagorean Association" soon came to prominence, leading Croton to a new political and military status. They governed the state so well that it became an Aristocracy ("Government of the Best") (This must also have been a little like the mafia!), which lasted twenty years. About 500 BC many leading Pythagoreans were murdered in a revolution, but Pythagoras himself escaped these "Anti-Pythagorean Uprisings." Almost all of the Pythagoreans had left Italy by 400 BC (McKirahan, p.79-80).

Though Pythagoras was mostly a politician, he was also a religious reformer who created a way of life for himself and his followers, sometimes for good or ill (the description of his "cult" is reminiscent of a group like the "scientologists.") His teachings were essentially a series of "sayings," which his followers would memorize and pass down. Plato describes Pythagoras in his *Republic*.

> "*Is Homer said to have been during his life a guide in education for people who delighted in associating with him and passed down to their followers a Homeric way of life? Pythagoras himself, was greatly admired for this, and his followers even nowadays name a way of life Pythagorean and are conspicuous among others*" –Plato, Republic 10 600a-b.

More than any other presocratic philosopher he remains the subject of many legends. In one of these he is credited with biting the head off a snake, in another he is hailed by a river, and in yet another he is credited with making predictions! The neo-Pythagoreans of the mid-first century BC emphasized religious superstition and numerology. His followers who referred to themselves as "Akousmatikoi" (a Greek word meaning something like "thing heard") accepted his "sayings" on simply hearing them, while the "Mathematikoi" represented the scientific side of his religion, reflecting number over matter and preferring mathematical accounts of phenomena, over descriptions of constituents, and proofs begin to take precedence over "stories" (McKirahan, p.79, 83, 89).

His ideas about the generation of the physical world discuss geometry: points, dyads, lines, plane figures, space volumes, solid figures, and these are composed of "earth, air, fire and water."

> "*From the unit and the indefinite dyad spring numbers; from numbers, points; from points, lines; from lines, plane figures; from plane figures, solid figures; from solid figures, sensible bodies, the elements of which are four: fire, water, earth, and air; these elements interchange and turn into one another completely, and combine to produce a universe [Kosmos] animate, intelligent, spherical, with the earth at its center, the earth itself being spherical, and inhabited round about*" – Alexander Polyhistor,

> "*Pythagorean Notebooks*" quoted in Diogenes Laertius, "*Lives of the Philosophers.*"

> "*There are five solid figures called the mathematical solids, Pythagoras says that earth is made from the cube, fire from the pyramid, air from the octahedron, water from the icosahedron, and from the dodecahedron is made the sphere of the whole.*" – Aetius

> "*The Pythagoreans also said that void exists, and enters the universe from the unlimited breath, the universe being supposed in fact to inhale the void, which distinguishes things that are next to each other. This happens first in numbers; the void divides their nature.*" – Aristotle, Physics

This idea of a "*breath*" surrounding the Kosmos recalls Anaximenes and the way he described the earth as surrounded by air, the different shapes define the various "elements." This is a very different cosmogony than the earlier "Ionians."

Parmenides of Elea (b. c.515 BC) (non-Ionian):

Zeno's teacher, he is the first philosopher to introduce "deductive arguments" acknowledging their "compelling force." He made use of this new tool to raise basic questions; "What conditions must existing things satisfy?" or "Is reality what our senses tell us it is?" "How can we tell?"

For Parmenides, the world revealed to our senses is an illusion, there is but "one thing," and it doesn't change or move. It didn't come into existence and will not cease to exist. He is the first philosopher to undertake the philosophical analysis of "being" and of "coming into being," "change," "motion," "time and space."

Further, he was the first philosopher to meld these ideas into "*metaphysics*" (McKirahan, p.157-9) He wrote a famous poem in dactylic hexameter (like Homer and Hesiod), which was copied by Simplicius into one of his own works. The poem has three parts: a) Prologue: In which the goddess announces that she will tell Parmenides two things: First; "The unshaken heart is persuasive (or, well rounded) truth, and Second; "the opinions of mortals in which there is no reliance," taken together these two sections of the poem occupy the last section which is entitled "The Way of Truth and The Way of Mortal Opinions."

This last section establishes Parmenides importance as a philosopher (McKirahan, p. 160-178).

Zeno of Elea (b. 490 BC) (non-Ionian):

He defended the ideas of his teacher Parmenides along with his contemporary: Melissus.
Our best source of information about Zeno, his age, his relationship to Parmenides, and the purpose and nature of his work come from Plato (McKirahan, p.179).

His famous "*Arguments against Motion*" are remembered by people today as his famous "paradoxes:" The Dichotomy and the Stadium" (Achilles and the Tortoise), the "Arrow" (motion doesn't exist), "Place is in place," and "the millet seed" (McKirahan, p.179-193)

The first argument is that there is no motion because that which is moving must reach the midpoint before the end. The second is the one called "Achilles" and this describes how by running on a curve approaching an "asymptotic" finish line, even the mighty Achilles is robbed of his victory by the "geometric progression" (an infinite number of half-way points) and the fact he can never cross the finish line before (his slower moving rival) the Tortoise. In his third argument, Zeno makes a mistake in reasoning and suggests (quoted by Aristotle):

> "Everything is always at rest when it occupies a space equal to itself" and "what is moving is always "in the now," the moving arrow is motionless"- Aristotle, Physics

The fifth argument is called "the millet seed:"

> "Does a single millet seed make a noise when it falls, or one ten-thousandth of a millet seed? Isn't the ratio between the bushel of millet seeds and the millet seed, or even the one ten-thousandth of a millet seed, the same? So won't there be the same ratios of their sounds to one another? For as the things that make the noise (millet seeds in varying amounts), so are their noises (sound in varying amounts). Therefore, the millet seed does, in fact, make a noise"

> -Simplicius, Commentary on Aristotle's "*Physics*"

Empedocles of Acragas (f. 460 BC) (non-Ionian):

A contemporary of Zeno and Melissus and a generation younger than Parmenides, Empedocles came from Acragas, a Greek city in Sicily, which held to the Pythagorean intellectual traditions of western Greece.

He visited Thuri around 444 BC and may have met the historian Herodotus, as well as Protagoras, the Sophist. A political activist, he helped suppress an oligarchic regime, despite his membership in the Aristocracy and also fought for democracy (McKirahan, p.255-259). Later he was exiled to and may have died in Peloponese, but other accounts of his death are more sensational and have him jumping into the crater of Mt. Etna, thus confirming that he had become a god! But these stories are thought only to reveal his sense of "ego," and "showmanship."

He was hailed as "*someone who had been sent by god,*" when he helped the people of the neighboring city of Selinus, by freeing them from a pestilence caused by a polluted river. He did this by diverting two nearby streams at his own expense, thereby flushing out the unhealthy stream (Mckirahan, p.255).

Both a Physician and a Magician, which were not distinct professions in his day! He is supposed to have worn long hair and held to a grave demeanor, but was also a philosopher and cosmologist (Kosmos).

He preached a religion claiming a fall from a state of original purity and grace, and also of ultimate redemption. His writings include poems in dactylic hexameter in which he exhorts the reader to save his soul; he composed tragedies, a historical poem about the Persian War, and another poem for Apollo, medical writings, and philosophical poetry (McKirahan, p.255-259).

In his writings on physical principles he identifies two chief characters and processes of the Kosmos:

First; four elements: fire, air, water, earth, and Second; two sources of change: love and strife.

The elements: fire, air, water, and earth are eternal and fixed in quantity. They do not change into one another. They come into being, are destroyed, or undergo change in their basic qualities. They are also named for the gods:

Fire – Zeus (also "flame," "Sun"- Helios, "Shining one"- Elektor)

Air – Hera (also "Heaven" – Ouranos or Aither)

Water- Sicilian Goddess Nestis (or "Rain"- Pontos or "sea"- thalassa)

Earth- Aidonus (also Gaia, or Aia "earth")

The two sources of change: Love & Strife cause the reciprocal processes of unification and separation.

Are these like Forces? He describes them as material entities with spatial locations. Also, Love & Strife occupy different parts of the Kosmos and at different stages of its existence. The "Elements and Strife" are first mixed and later separated by "Love," and likewise, the elements remain separate from one another because of the continued presence of "Strife" (McKirahan, p.269-278).

Anaxagoaras of Clazemonae (500-428 BC) (return to Ionia):

His philosophy marks a return to the philosophical and scientific interests of the early Milesians, though he was also deeply influenced by the "Eleatic Philosophers." His most famous accomplishment was predicting that a meteorite would hit the Earth at Aegospotami (a city on the Gallipoli Pennisula) (McKirahan, p.200-1).

He moved to Athens and lived there forty years. While living there, he was an associate of <u>Pericles</u>, the great Athenian statesman, however, this association eventually led to his being tried and convicted for "Impiety" because he believed the sun to be not a "god" but a "fiery stone."

Because of this event, he became the first philosopher to be convicted in Athens. Later Socrates would receive a death sentence and Aristotle would flee for his life! (McKirahan, p.200-1).

Anaxagoras would find himself exiled from Athens and instead lived out his life in Ionia, in the city of Lampsacus, near Troy. He is presented here out of chronological order. He was actually younger than Parmenides, but somewhat older than both Zeno and Empedocles who were both born c.490 BC (McKirahan, p.200-1).

Anaxagoras' accounts of the origin of the Kosmos and of the nature of mind and matter are intimately related and inseparable from one another. They are founded on five kinds of entities and six basic principles (Mckirahan, p.203-227).
The Five kinds of Entities are: 1) Ordinary macroscopic objects, 2) Basic Things, 3) Portions, 4) Seeds,
5) Mind

The six principles are the following:

P1 – There is no coming to be or perishing.
P2 – There are many (perhaps unlimitedly many) different types of Basic Things.
P3 – There is a Portion of everything in everything.
P4 – Each thing is most plainly those things of which it contains the largest portions.
P5 – There are no smallest portions.
P6 – Mind (Nous) is unmixed with other things and has the following functions: a) it knows all things, b) it rules all things, c) it sets all things in order, and d) it causes motion.

It remains to examine his cosmology, which after all enabled him to make his meteorite prediction in 467 BC:

1) Believed all Heavenly Bodies are made of "Stone."
2) "The earthen bodies held aloft by the cosmic vortex can sometimes slip and fall to earth."

This reference credited to Anaxagoras led to his receiving the credit for the famous "meteorite" (Today, we know that about 15 tons of these rocks fall in everyday! So of course he got it right!)

Leucippus of Miletus (f. 440 BC) & Democritus of Abdera (b. 460BC) (The Atomists)

The most ambitious challenge to the Eleatic Philosophers came from the "Atomic Theory" which was first developed by Leucippus and later refined by Democritus.

Democritus was Leucippus' student and it is likely that Leucippus formed the atomic theory in the decade 440-430 BC and wrote two books: "*The Great World System*," and "*On Mind.*" Democritus' birth date is inferred from his own statement that he "was young in the old age of Anaxagoras (c.500 BC). He is credited with further developing the atomic theory in his adult years, showing that he had accepted his teacher's beliefs, wheras Leucippus was a shadowy character who had lived in many places including: Miletus, Elea, and Abdera. Though this just reinforces, that his philosophy was distinctly "Ionian." He was well aware of the "Eleatic Challenge" to the Ionian School of thought (McKirahan, p.304-4).

Diogenes Laertius reports that Anaxagoras was forty years older than Democritus. It has been noted that since Democritus was only ten years younger than Socrates, designating him in history as a presocratic isn't correct, strictly speaking. He lived to a ripe old age (over 100!) and lived well into Plato's lifetime and even into Aristotle's. He was born in the Thracian mainland in a remote Greek city of Abdera. He traveled extensively in non-Greek lands for study and research, he is well known to us through the extensive writings he left and are still available on a number of topics including: ethics, natural philosophy, mathematics, music, and various technical subjects including medicine and military strategy. He is sometimes remembered as the "laughing philosopher" for his sense of humor. More fragments from Democritus survive than for any other presocratic philosopher. Most of these writings are on ethics, however, and have nothing to do with atomic theory (McKirahan, p.303-4).

Attempts have been made to distinguish between Leucippus and Democritus, but have failed because of the close relationship between them and their accomplishment. Leucippus, essentially proposed a model for the cosmology of the universe (as a response to the "Eleatic Philosophers) and Democritus did what no one had ever done before, he accepted his teacher's theory and went on to explain a wide range of natural phenomena based on that theory, while along the way he realized the need for a more rigorous mathematical basis for his physical theory and the need for an appropriate theory of knowledge to go with it (McKirahan, p.303-4).

Democritus tried to establish a thoroughly atomistic account of all aspects of the world. Not one of his works survived because he wrote in Abdera instead of Athens, and also because his theory was not well liked by either Plato or Aristotle (and this led to his getting overlooked!).

In his Atomic Theory there are two types of elements: Atoms and Void

Atoms are indivisible (Atomos – "*uncuttable*") building blocks of everything, which are too small to be seen, which move in the void and combine to form compounds, some of which are large enough to be perceived with the naked eye. Atoms are described as "Full," "Solid," "Compact," "What is," "Being," while Void is described as "Empty," "Rare," "Unlimited," etc. The Void enables atoms to move and to preserve the uniqueness of their identity. Infinitely many atoms are all in motion in the infinite void. As an atom moves it may meet with other atoms of the same kind or of different kinds. Such collisions can result in the atoms rebounding away from one another or in their coming together to form compounds (McKirahan, p.304-317)

Conclusion:

In summing up the contributions of the pre-Socratic philosophers we are left with many controversies and unanswered questions, especially with regard to <u>Thales</u>, the founding member of the Ionian school, and about whom so little is really known. Much of what is credited to Thales is either borrowed from other cultures, as in the case of the Babylonian astronomy used for his eclipse prediction, or his geometry, which probably was based on Egyptian learning, or even his belief in water as a material from which all things are made. This also reflects the culture and religious traditions of Egypt. Many of his accomplishments, such as the authorship of a certain "Nautical Star Guide" were attributed to him because he had a kind of "rock star status" because he was one of the "seven sages" and because it was the local custom to honor the "oldest" and "wisest man," so of course Thales was him! He was the oldest and the wisest! Even in matters where he was being scientific, he still ascribed natural phenomena like "magnetism" to the piece of amber having a "Soul," thus showing that he was not completely over invoking the gods to explain natural phenomena. For him, as a transitional thinker, the amber was still "full of gods!" Finally, there is the incomplete and fragmentary nature of what is known about him. Most of what we know about him is only reflected in the writings of other philosophers and most notably- <u>Plato</u> and <u>Aristotle</u>.

To comment on the contributions of the other pre-Socratics it deserves to be said that they carried forward the "Critical Examination" of nature begun by their founder and set themselves apart, as a major turning point in the history of human writing and thought.

Their real accomplishments and indeed the accomplishments of the whole Ionian school amount to just these major innovations: 1) Critical Examination (Observation & Rationality) and 2) Critical Attitude (refuting each previous generations ideas using "Arguments"). It isn't important whether they held water, fire, air, or earth, to be the original primordial stuff out of which everything is created, because each generation contradicted the ideas held by the previous one! It is not until <u>Parmenides</u> that deductive arguments are used to explain such difficult philosophical questions as "change" and "motion.'

This new tool led <u>Zeno</u> to conclude that motion doesn't exist and he proposed his "*arguments against motion*" to explain it. <u>Empedocles</u> invoked a cosmic interplay between the four elements, love and strife to explain motion and the apparent diversity of substances. Finally, The Atomists conclude that the kosmos is void filled with large numbers tiny, uncutable atoms, which collide to form compounds some of which are large enough to be seen with the naked eye.

Even though the historical record is fragmentary and there are lingering doubts about their "originality," the presocratic philosophers deserve a special place in history and to be recognized as having established "a western foundation" if not necessarily the "original foundation" for western science.

And even if their accomplishments seem to be marred or flawed, it is still a remarkable story of how the early Greeks, with nothing but their intellects and imported learning, established the critical framework for the philosophy of western science, which has lasted down through all the subsequent centuries. Some of their ideas still manage to sound surprisingly modern to us, or is it that living in a culture, which has been as strongly influenced by their philosophical tradition as ours has been, that we owe many of our so-called "modern" ideas and intellectual concepts to this particular "era" of thinkers, if not always necessarily to the individuals (both real and imagined) credited by history with their discovery or creation.

Galileo and Experimentation

The following is an excerpt from "The Copernican Revolution" by the author David Tracy prepared for a class at Cal-State L.A.

By the middle ages it was the "Eudoxan version" of the heavens, with perfect crystalline spheres, relayed in the writings of <u>Aristotle</u>, which became the officially held view of the Roman Catholic Church, during the life of *Galileo Galilei*, whose simple observations through a new device called the "telescope," first called into question the accepted wisdom of these "perfect concentric crystalline spheres" in the heavens. His classic works "*The Starry Messenger*," "*Letters on Sunspots*, and especially his "*Dialog on the three world systems*," created considerable "ill will" between himself and *Pope Urban VIII*.

This conflict became one of the classic confrontations between science and religion, and Galileo paid a heavy price for his "academic freedom!" Even though his observations through the telescope would eventually be proven correct (even by the church itself, since so many priests were also interested in the astronomy of the day), he was still sentenced by the "Inquisition" to *house arrest* for the rest of his natural life for "holding and defending" the Heliocentric Copernican system in defiance of the orders of *Cardinal Bellarmine* his Inquisitioner!

Galileo Galilei (1564-1642)

Our treatment of the life and astronomical achievements of Galileo will be in two parts. The reason for this is that unlike the other astronomers of the so called,
"Copernican Revolution" (such as <u>Nicholas Copernicus</u> himself, who authored the very influential "*Revolutionibus de Orbium Coelestium*"), he didn't actually create any new theories to do with planetary motion. That is to say, his greatest achievements were not in the form of "orbital calculations" like <u>Johannes Kepler</u>, but in performing observations using a new device —the "Galilean Telescope," and writing about what he observed. The other thing for which he is remembered in this history is being put on trial before the Holy Office of the Inquisition for heresy, subsequent to his long held private conversion to "*Copernicanism*," and writing the "*Dialogue*."

This material is very different from the other sections and stories surrounding it in this paper, which are about ancient Greek philosophers, and their search for a primordial stuff out of which all things were made. By the middle ages certain men, like Kepler, were trying other geometries and computations to describe the motion of the planets and their search must be cast against the "spirit of the times" and the absolute dominance of Aristotle's views over all intellectual activity, but especially the church and its teachings about the heavens.

His Early Life:

Galileo was born in Pisa in 1564, a part of the Grand Duchy of Tuscany (Italy), which in this time was ruled by <u>Cosimo de Medici</u>. His father <u>Vincenzio Galilei</u> was an accomplished musician, composer and mathematician, who sent his son off to study medicine at the University of Pisa, but instead he became interested in mathematics (Goodman, p.92).

At age twenty five, he was appointed professor of mathematics at the University of Pisa with the help of a family friend and nobleman <u>Guidobaldo dai Monte</u>.

It was during this period when he is rumored to have dropped balls off of the famous Leaning Tower in the presence of university professors, to falsify Aristotle's theory that;
"*Heavy bodies of different weight fall at different rates, with velocities proportional to their weights.*" However, this legend has since been discredited (Goodman, p.93).

Perhaps motivated by the university's resentment of Galileo's opposition to the established doctrines of Galen and Aristotle, he left Pisa in 1592 to fill a vacant chair in the mathematics department at the University of Padua in the Republic of Venice. It is here that Galileo produced his most significant work, the kinematics that would transform the physical sciences (Goodman, p.93).

His lectures at Padua were all conventional and based on Ptolemy, but in private he was already converted to the new astronomy of Copernicus. In His little book "*The Starry Messenger*" he described what he had seen through the telescope, a very new view of the heavens, which contradicted the widely held teachings of Aristotle.

Aristotelian Dominance:

Aristotle had become more than just the dominant voice of the age by the time of Galileo (14th &15th centuries). It had become the very language of the church and of every educated person of the day because of the church's pervasive influence over education. The only way one could receive an education was to be taught by clerics (who themselves had been taught by clerics) steeped in the Aristotelian tradition.

Aristotle was an established authority whose writings formed the basis for logic, natural philosophy, and most important provided the framework which reinforced the church's dogma (Goodman, p.93)

Many early church writers, such as; Albertus Magnus (c.1200-1280) and Thomas Aquinas (c.1225-1274) had shown how the teachings of Aristotle could be joined with Christian Doctrine, a relationship, which has lasted ever since.

According to Aristotle the universe was a sphere of limited size, with the Earth, motionless, at its center since this was its "natural place." The earth was then surrounded by a series of concentric spheres each corresponding to one of the natural elements (water, air, & fire). The sphere of fire was surrounded by a succession of seven solid crystalline spheres carrying the moon, Mercury, Venus, the Sun, Mars, Jupiter and Saturn. The next sphere beyond them contained the fixed stars and beyond them was nothing (Goodman, p.94). The sphere of the Moon was a fundamental boundary for Aristotle because two distinct sets of laws operated in the universe: 1) one for the region beneath the moon, 2) and another set for the spheres beyond. This "Celestial Region" had no imperfections, while the sub-lunar "Terrestrial Region" was associated with imperfection. Aristotle had also taught that a "Prime mover" (i.e. God) was the ultimate source of motion in the universe (Goodman, p.95).

So long as Aristotle's cosmology held sway, it presented an impenetrable barrier to new ideas and especially to Copernicanism.

Galileo's Telescope:

Some versions of history try to give Galileo credit for inventing the telescope, but this just isn't true!

What he did do was to improve an existing design. In 1609 news reached him of a new instrument, which had been invented in the Netherlands, which made distant things appear closer. In his first attempt he improved the magnification from 4X to 10X, by simply placing a concave lens behind a convex lens (the basic design for the Galilean Telescope which bears his name). By his third telescope, he had made a telescope capable of 1000X magnification (Goodman, p.95).

Another accomplishment of Galileo was to realize before anyone else in Italy just what the advantages of this new instrument would be for armies on land and navies at sea, so he took his new device to the Venetian Senate to show them what it could do. The other thing, which it could do, was less obvious in the early 17th century, and that was looking at the heavens. Galileo was the first to turn his telescope towards the heavens, his astonishing observations, which in them selves were discoveries, are published in his little book "*The Starry Messenger*" (Goodman, p.95).

The Starry Messenger:

He noted that the surface of the moon appeared "pitted with craters" and that there were ""Mountain peaks lit by sunlight, while large areas remained in darkness." Galileo thought incorrectly that the dark areas on the moons face were "Seas" (hence the term "Mare" for these areas) and he calculated that some of the mountains on the moon must be about four miles high, which is now known to have been a good estimate (Goodman, p.95).

These observations, however, conflicted with the fundamental Aristotelian view in which heavenly bodies were seen as being perfect.

He went on to give new counts as to the number of stars and reasoned correctly that the naked eye overestimates their size (that they are vastly more distant and this explains why they appear as points). Finally, he turned his telescope on Jupiter and observed four tiny nearby stars. On the following several nights he recorded observations for their movements and concluded that they were orbiting around Jupiter itself. Galileo named these moons of Jupiter the "Medicean Stars" in honor of his benefactor Cosimo II de Medici (who had been his pupil) now Grand Duke of Tuscany (Goodman, p.97).

It has been suggested that Galileo did this to court favor with the Tuscan ruler, for just a few months later after the publication he was appointed chief mathematician and philosopher to the Grand Duke and the head of the mathematics department at University of Pisa.

These observations hadn't proven the new theory proposed by Copernicus, but they had added considerably to there plausibility and hundreds of copies of The Starry Messenger were soon printed. Kepler himself, enthusiastically received the work (Goodman, p.99), but a more serious difficulty arose confirming the observations. In Rome, <u>Father Clavius</u>, the Jesuit authority on astronomical matters was soon able to confirm these observations. As a result, Galileo was invited to Rome for a cordial interview with

<u>Pope Paul V.</u> During this visit Galileo was elected into the "Accademia die Lincei" (a special society within the church dedicated to the study of natural philosophy). The trip had been a great success; he had even been introduced to <u>Cardinal Robert Bellarmine</u>, who would later become his Inquisitor!

Belarmine was the leading theologian of the Catholic Church and the Guardian of orthodoxy! Written as a set a personal letters between himself and one Mark Wesler (an Augsburg merchant with an interest in natural philosophy), Galileo's Letters on Sunspots were received by the Accademia dei Lincei without incident except for Jesuit astronomer <u>Father Christopher Scheiner</u>, who also claimed credit for the discovery of sunspots. This would mark the end of Galileo's work in astronomy, but the trouble over it was just beginning.

The Trial of Galileo before the Inquisition:

The real source of Galileo's dispute with the Church was a discussion at the dinner table of the <u>Grand Duke Cosimo II</u> in December 1613 (Goodman, p.103)

The guests included Boscaglia, a professor of philosophy at the University of Pisa, and <u>Benedetto Castelli,</u> a Benedictine monk and former pupil of Galileo's, who had just been appointed to the chair of mathematics at Pisa with specific instructions "not to teach the motion of the earth!" (Goodman, p.103)

The conversation at dinner turned on the topics of; the university, the telescope and astronomical observations.

The <u>Grand Duchess Christina</u>, the mother of the Grand Duke, asked questions about Jupiter's moons and assurance was given that they really existed (Goodman, p.103)

Boscaglia confided the Grand Duchess that while all of Galileo's observations were confirmed, his statements that the earth moved was incredible! (Chiefly because it contradicted Holy Scriptures) (Goodman, p.103).

In the ensuing argument, one of the guests (Castelli) beat a hasty retreat only to be returned, so the Grand Duchess could quote scriptural passages against Castelli's view and in favor of Galileo. Castelli used theoretical arguments, which persuaded the Grand Duke and while the Grand Duchess continued the dispute, Catelli continued to use this as pretext for hearing his replies. All throughout Boscaglia remained silent (Goodman, p.104).

Castelli described all that had taken place in a letter to Galileo.
A week later Galileo sent a very long letter in reply expounding his views on the relationship between religion and science.
Castelli circulated copies of this letter amongst his friends.

A copy of the letter made it to Father Lorini, a Dominican who had spoken against "Impertnicus" The contents of the letter shocked him and he at once contacted the Holy Office of the Inquisition in Rome, enclosing a version of the letter which was not wholly faithful to the original (Goodman, p.104)

Galileo was anxious to have an original copy sent to Rome and asked if a friend of his; <u>Archbishop Piero Dini</u>, would show a copy of it to Cardinal Robert Bellarmine, but nothing came over Galileo's views on science and the scriptures. The authorities in Rome could find nothing heretical about them (Goodman, p.107).

Later in the nineteenth century, Galileo's own enlightened opinions would be endorsed by the church, but in his own time he was under the scrutiny of the church and the Council of Trent had prohibited expounding the Bible contrary to the common agreement of the holy fathers. All of the writers of that time expounded the geocentric Aristotelian cosmos. The church authorities asked Galileo (Goodman, p.108);

"What's your proof that the earth is not in the center, but orbits third away from a central sun?"

To which they offered from the book of "Ecclesiastes"1:5; (Goodman, p.108)
" The sun also riseth and the sun goeth down, and hasteth to the place where he ariseth"

These words spoken by Solomon (who had wisdom from God) according to the "appearances of the sun and earth" (that is it seems to us as if the sun goes around the earth!)

On December 11, 1615 Galileo arrived in Rome to defend his views on the Copernican system (that it was physically true).
Pope Paul V intervened with a request for an "official statement" on the motion of the earth and the stability of the sun. A report was issued by the theological experts from the Congregation of the Index (Goodman, p.109), on Feb 24 1616, the theologians concluded that;
The proposition that the sun was stationary at the center of the universe was declared "foolish and absurd," philosophically and heretical, because it contradicted the Holy Scriptures.
As a result certain books were banned and others modified to reflect the order.

Cardinal Bellarmine called Galileo before the Inquisition for an interview which resulted in a certificate being given to Galileo demanding that he "not hold or defend" the Copernican theory (Goodman, p.109-110);

"We Roberto Cardinal Bellarmine, having heard that it is calumniously reported that Signor Galileo Galiliei has in our hand abjured and has also been punished with salutary penance, and being requested to state the truth as to this, declare the said Signor Galileo has not abjured, either in our hand, or the hand of any other person here in Rome, or anywhere else, so far as we know, any opinion held by him; neither has any salutary penance been imposed on him; but that only the declaration of the Holy Father and published by the Sacred Congregation of the Index has been notified to him, wherein it is set forth that the doctrine attributed to Copernicus, that the earth moves around the Sun and that the Sun is stationary in the center of the world and does not move from east to west, is contrary to the Holy Scriptures and therefore cannot be defended or held. In witness whereof we have written and subscribed these presents with our hands this twenty-sixth day of May, 1616"

Galileo kept a copy of this certificate, but another document giving a very different account of the interview appeared in the Vatican file of the event. It would be the discrepancy between these two documents, which would cost Galileo dearly. A new Pope Urban VIII would replace Pope Paul V and even though they had been friends, he would punish Galileo by putting him under house arrest for the rest of his life. One reason had been that he had made fun of the Pope himself and his adherence to the Aristotelian cosmology, by putting the Pope's own words in the mouths of one of the characters in a play called the "Dialogue." Written in Italian, the play takes the form of a "Socratic dialogue" between three friends in the city Venice, who have conversations on each of four days about various topics in motion. One was the character of *Giovanni Sagredo*, a Venetian nobleman, another was *Filippo Salviati*, the last and final character is *Simplicio*, from whose mouth the Pope's own views come in one of the closing speeches (and who for most of the play represents the philosopher Simplicius) (Goodman, p.110-116, & 119).

Only during the reign of *Pope John Paul II* did the church officially acquit Galileo for his heresy! During his life there were even those superstitious clerics who advised people not to look in the telescope, because they might become possessed by "evil spirits." Such notions in today's world seem ridiculous. Or do they? Have ordinary people become more open minded since the days of the "Inquisition?"

In truth, Galileo's experimenting proved to be a crucial departure from what had been done before! Further, not all of his experiments were intended to be "pure science," but rather were often for "purely applied purposes" such as the "pulley problems" found in most high school physics books, which owe their whole presence there to Galileo's search for refined methods by which to ship "Venetian Glass" to every important port in Europe for his royal benefactor; *Cosimo de Medici II,* in whose honour he named the four largest moons of Jupiter as the "*Medecean Stars,*" thus showing that he was both scientifically curious as well as seriously pragmatic.

Without his pioneering work in the area of "accelerated motion" much of the work of *Sir Isaac Newton* might have languished due to a "lack of attention."

As it was, Galileo begged Newton to publish his major work on the topic of theoretical mechanics; "*Philosphiae Naturalis Principia Mathematica,*" so that he could sight the work as a relevant source for the support of his own work!

No other thinker in history had ever sought to put *nature* to the *test* and show that there were solid principles which underlie the way things move and behave, that such principles could be explored and understood through experiments, and that doing so would hold great economic and industrial benefits!

Though Galileo paid for the intellectual independence of his science with the loss of his personal independence, his critics and detractors are now lost to history, overcome by the scorn heaped upon them by the broader, more liberal, and *open minded* sweep of histories written by "new thinkers," people who now realize just how wrong the church was to act in this way!

Galileo was even able to "smuggle" one of his later works out of Italy for publication in Holland because of his close commercial ties with that nation and the help of his daughter, a nun named "Maria Celeste."

Sir Isaac Newton: Theoretical Mechanics

Though Born prematurely at Wollsthorpe, a hamlet near Grantham, in Lincolnshire, Sir Isaac Newton (1642-1727AD) was destined to become one of the greatest mathematicians of all time. After his father's death in his early life, he was left in the care of his grandmother, during which time his mother remarried.

After his new stepfather died also, his mother decided that he should become a farmer and learn to manage their estates, however, the young Newton had no such interests! In 1661 he entered Trinity College, Cambridge, as a "*subsizer,*" that is to say an impoverished undergraduate student who performed menial tasks in exchange for money to put himself through school. Graduating in 1665, he had already developed a new mathematical method based on the use of something he called "Fluxions" which would anticipate his development of the Calculus.

When an outbreak of "the plague" forced the closure of Cambridge, Newton fled to the country side to escape danger. It was during this period that he consolidated his mathematical work on the Calculus and motion, as well as his optical work on the "theory of colour."

It was also during this time that he realized that the planets were held in their orbits by some force emanating from the sun, which diminished inversely with distance from the sun- "An Inverse Square Law," yet he published nothing during this period!

He returned to Cambridge in 1667 and found himself elected to a "Fellowship" (what we would call a professorship) at Trinity, and within two years he became the "Lucasian professor of mathematics (what we would describe today as a research fellowship). Along with his new theoretical developments, in 1671 he built the first "reflecting telescope," which used a parboloidal mirror instead of lenses to focus an image within an eyepiece. Presenting a copy of his new design to the Royal Society, this bold original work refuted the widely held belief of the time that no

"Achromatic" (without chromatic aberration) telescope could be made, unlike the refracting telescopes of his time which suffered greatly from this chromatic effect (stray color, or dispersion). He soon submitted his treatise called the "*Theory of Light and Colour,*" which led to such a great controversy for its author, he didn't publish anything else until 1675. In that year he presented two papers; the first was concerned about the colours seen in thin films of oil, and the other introduced his idea that light was composed of tiny particles of light called "corpuslcles."
These papers were again met with controversy and he was quite sensitive to criticism.
It was not until 1679 that he again resumed writing letters, this time to *Robert Hooke,* who though another of Newton's critics was also now the Honorary Secretary of the Royal Society. This exchange caused Newton to re-think his work on planetary orbits, but it was not until 1684, with a visit from *Edmond Halley,* that he again considered making his ideas public.
Halley, who had been in contact with Hooke and an architect friend named *Christopher Wren,* who had an interest in the orbits of planets, explained to Newton the no one had yet been able to show in the form of a mathematical proof, that that assuming an inverse square law would lead to elliptical orbits! Newton took up the challenge and proved this to be the case in a small work he named "*De Motu*" ("On Motion"). Halley himself realized the importance of the work when he received it and notified the Royal Society with a request for publication, however, financial problems forced him to publish the book himself at his own expense!
Only the cool head and diplomatic skills of halley saved the work from obscurity, published with the title "*Philosphiae Naturalis Principia Mathematica,*" it covered the whole question of the motion of heavenly bodies, incorporating the "Laws of motion" and the new "Universal Gravitation."
Exhausted from his intellectual efforts, or perhaps fearing another critical confrontation with Hooke, it seems that after the publication of the "Principia" that Newton suffered a nervous breakdown! This caused him to withhold the publication of his book "*Optiks*" and his acceptance of the presidency of the Rotal Society until after the death of his archrival Hooke.
During the years that followed he devoted himself to "religious studies" and "alchemy."
In 1696 he was appointed "Warden of the Mint" and later "Master of the Mint," after he had overseen the recoinage for all of England! For his outstanding effort he was rewarded with a "knighthood" in 1705 (this is why he is called "Sir Isaac"). By the end of his life he was recognized internationally as the greatest scientist of his day. He died March 20, 1727 in London and was buried at Westminster Abbey following a state funeral (a great honour for an Englishman!).

In the next section we will begin developing our understanding of terms and definitions from "Newtonian mechanics" so that we can do "creative experimenting" a little later in the book.

Albert Einstein and Curved Space-time:
(The following is an excerpt from a paper by the Author (David Tracy) entitled
"A commentary on the "Poetry of the Universe", by Robert Osserman" and was prepared for a college class at Cal-Sate Long Beach).

I read Robert Osserman's book "*Poetry of the Universe*" and found it a very nice introduction to some of the more "abstract ideas" found in Modern Geometry. What this book is all about was the "inevitability" of the discovery of "non-Euclidean Geometry," the scramble among the mathematicians of the nineteenth century to claim credit for It, and how men like Albert Einstein employed it in their own work later. His theories of "*Special*" and "*General Relativity*" and the very fact that the "moving" universe couldn't be adequately explained with less than a "four-dimensional curved space," and the contributions from many generations of "Geometers" both known and unknown, that led to those developments. The book begins by discussing the work of earlier Greek and Arabic geometers as a pretext for the later and more sophisticated discussion of "Geodesy" and its relationship to "Mapping" during the renaissance.
 In many ways the "Shape of the Curved Earth" is what made the development of the "new" non-Euclidean geometry so inevitable. Trying to represent a three-dimensional earth in two-dimensional form, eventually forced certain developments to take place. For example, longitudinal lines could not really be described as being "parallel" in the "Euclidian sense," because they converge at the poles and this fact led many mathematicians of the nineteenth century to create new forms of geometry based on "new" assumptions about parallel lines.

In order to show how all of these developments came about, I need to outline both the course of 4000 years of both geometry and it's history as described in this book. This will serve to highlight the "historical developments" and show their connection to the "geometrical ones."

Through out the paper the "names of the mathematicians" will be "underlined" and the "titles" of any famous "classic" book references will be "italicized." I will do this to preserve the developmental flow of Osserman's book and to bring out the historical figures in the story, as well as some of the "classic texts" he describes.

Modern theories don't just come out of nowhere and some of the most abstract ones actually had their origins in the much earlier works of unknown Arabic and Greek mathematicians. In the end we will see that it is from the work of Georg Riemann, that Albert Einstein first learned of the "geometrical ideas" he would later remold and include in both his "Special and General theories of Relativity."

It could be said that the earlier, and less well known, work of Riemann on "Curved Spaces" influenced the later and more celebrated work of Einstein.

To begin our discussion about the history of geometry, we first need to begin in a time when things were much "simpler." Back in ancient times, when men needing to construct buildings, first learned to create "maps representing the earth," and were, for the first time, compelled to "reason out on paper" the first principles of two and three-dimensional geometry.

"Geometry" literally means, "Measuring the Earth" and the book plays on this idea. It begins by retelling the story of Pythagoras in ancient Greece and the discovery of the "3-4-5" Triangle. The book confesses that this problem may have been solved by earlier Arabic geometers, as were a number of other developments, most notably the "Pythagorean Theorem," which had been worked out, at least practically, by the builders of the Pyramids, who were probably from Babylon (p.4).

What sets Pythagoras apart from the other geometers, and the reason that this very famous theorem is named for him, is that no one before him seems to have bothered to invent the notion of the "*Proof*." The book goes on to admit that it was probably not proven by Pythagoras either, but more likely by one of his "*Pythagorean followers*" a century later. Osserman next brings up Euclid, the most famous of all Greek mathematicians, who was born over 200 years after Pythagoras. During the time between them, geometry develops along two parallel tracks; one devoted to "shapes" triangles, rectangles and bounded curves; The other, to the development of "methods of proof," for the universality of the discoveries by "deductive reasoning." This led to discoveries that otherwise wouldn't have been found except by direct observations. Indeed, by the time Euclid comes on the scene a vast body of geometric knowledge had accumulated from earlier times.

The details of Euclid's life remain a mystery, shrouded in obscurity. What is known, is that he lived and worked in Alexandria around the year 300 B.C. and unlike Pythagoras, left a large body of written work, which survives to this day and have become the "models" for all mathematical research (p.5). The work for which he is best known is "*The Elements*." It was a compendium of 13 books, five of which were about geometry of 2-dimensional shapes, 3 to 3-dimensional geometry and the rest to other subjects.

"The "*Elements*" had an important effect on the psyche of western civilization. It grew to become *"a body of knowledge to be digested and retained by students for 2000 years."*

According to Mr. Osserman, the "*Elements*" had four distinct components. First, there is a sense of certainty. That in a world full of speculation and conjecture, the statements in the "*Elements*" were proven true beyond a doubt.

This is one of the major points of the book, that, even after 2000 years "no one" has ever found a "mistake" in the "*Elements*" (p.6, and this not only established the 'tradition" in mathematics of the importance of always "getting the right answer," it is a great tribute to Euclid and all who preceded him, that they were able to do it all with nothing but paper and pencil!)

Second, is the power of the method. Starting from a few very simple "*assumptions*" Euclid derives a dazzling series of geometric consequences. Third, is the "ingenuity" of the proofs. Finally, there is an "aesthetic appeal" to the geometric shapes themselves apart from the lines of proof associated with them.

Of those shapes studied by Euclid, none are more beautiful or more useful in terms of "measuring the earth than the "Circle" (p.5-6).

Osserman conjectures that it was the circular shape of the "Horizon," which first gave men the impression of what the real "Shape of the Earth" could be. He further adds that it was the "circular motions of the night sky" which provided the first "concrete evidence" for the spherical shape of the earth. The more challenging problem was not the "shape" of the earth, but the "size" of the earth.

A clever solution to this problem was developed by Eratosthenes of Alexandria. His method consisted of three parts: 1) the fact that a special geometric relationship exists between the sun's shadow and the earth at Equinox (i.e. there are "Similar Triangles" that can be used to calculate the latitude), 2) The "distance" between Alexandria and the city of Syene (now called Aswan) was "known," and 3) the use of a simple device called a "Gnomon," essentially a "Stick of known height" which casts a shadow on level ground that can be measured (see appendix A). The "arctangent" of the ratio of the shadow's length and the height of the stick provides an "angle," which corresponds to the gnomons "latitude." The gnomon serves as a primitive calendar and can also measure two important days of the year: "the summer and winter solstices." It can also be used to measure the "angular distance of the sun" relative to the horizon p.11). The fortuitous properties of the location of Syene (Aswan) are what gave Eratosthenes the idea to try measuring the earth's diameter, since it lies almost due south of Alexandria and on one day of the year (Vernal Equinox), the sun's noon shadow is "non-existent," because the sun is directly overhead. Syene lies almost on the "Tropic of Cancer," a circle around the earth at 23.5 degrees "North" latitude (p.13). By combining these facts Eratosthenes was able to draw similar triangles and compute the "Circumference of the Earth." He found the angle to be 1/50th of the circumference of a circle (p.13). This meant that the circumference of the whole earth had to be 50 times the distance between Alexandria and Syene. Since that distance is roughly 500 miles this suggested that the circumference was about 25,000 miles around (p.14). His method had a number of inaccuracies (p.15). First, measuring the angle between the sun and the vertical angle could only be done approximately. Second, Syene (Aswan) is not perfectly due south of Alexandria. Third, it would have been difficult to get an accurate measure of the distance between them, since there is great uncertainty in the interpretation of ancient units of measure. Eratosthenes' estimate was the most famous measurement, but not, by far, the only measurement. Aristotle, among others in the last century, had reported estimates for how big the earth was. These and subsequent estimates of the earth's "size" were to play an important role during the "Age of Exploration."

Ptolemy lived in Alexandria about the time of the Roman Empire, and is remembered for his work in consolidating and extending the work of previous centuries. He is best known for his work "*The Almagest*" ("The Greatest"), the first definitive descriptive work on the heavens and earth. It became to Astronomy what Euclid's "*The Elements*" had been for Geometry, the definitive treatise "for over a thousand years" (this is a far more significant achievement than it sounds. No modern textbook can come close to this because they go out of style so fast! When was the last time you saw an astronomy book last for more than a decade before it became "dated"?). In the same way Ptolemy's book "*Geography*" became the standard reference for the subject of "describing the earth" (p20-21). In Roman times, geographical information was "scanty" and not to be taken too literally. Eratosthenes is credited with being one of the first to apply mathematical methods to "Map Making." His methods were later refined by the astronomer Hipparchus, who is credited with developing "Trigonometry," the systematic study of the relationships between the sides and angles in a triangle. After the fall of Rome, map making reverted back to its previous, more fanciful state, based on hearsay and superstition, rather than scientific methods. It was not until the thirteenth century that the "Geography" was again available, but only in the original Greek, a language that was not widely known. Another 200 years past before it was translated into Latin. The first printed copy dated to 1472. Christopher Columbus owned a copy dated 1479.

The "Spherical shape of the Earth" is described as an "accepted fact" in Ptolemy's "*Geography*." But some how "Medieval Europe" lost some of the knowledge gained by earlier cultures. Subsequent publications in that era, suggest that only the educated people knew the true shape of the earth, in the fifteenth century. One of the titles gives a clue as to how the people of that age thought about the earth's correct shape, it was entitled "*The Sphere*."

The book's author was "Sacrobosco," a latinized version of the English name "John of Holywood," who is known for writing many books at the end of the thirteenth century. One of the most successful history textbooks in history, "*The Sphere*" it remained in print for over 500 years! What he did was to take important references out of Ptolemy's "*The Almagest*" add some newer material and eliminate some technicalities,
"*To arrive at a more accessible description of the inner workings of the universe, as best understood at the time*"(p.23). As the title indicated the "Sphere" was the key to everything. The Earth was a sphere inside another sphere of "fixed stars," while the sun, moon, and planets revolved on intermediate spheres. It included a variant of Eratosthenes estimate of the circumference of the earth and a calculation of the earth's diameter using "22 / 7" as an estimate for "Pi." It was "Required Reading" for the "A.B.degree" in Vienna in 1389, as it was in at Oxford in 1409 and Erfurt (Germany) in 1422.

By time of Christopher Columbus the concept of a "Spherical Earth" was neither idiosyncratic nor controversial, many scholars accepted it. The real question was not the "Shape" of the earth, but rather, what was its true size? No one really knew, because no one had made the trip to find out! (p.25). Ptolemy's estimate of 20,000 miles for the circumference of the earth, though a bit "low," suggested the "feasibility" of such a voyage to Columbus (p.25). In addition to underestimating the circumference, Ptolemy over estimated the size of Asia, which also made the distance between the western tip of Europe and the eastern edge of Asia look shorter than it really was. To Columbus it looked to be well within the range of a ship's provisions.
In 1484, King John II of Portugal appointed a team of experts to review proposals for "maritime exploration" and to advise him as to their "feasibility." This team was well versed in all the scholarly work on geography and navigation of the day, as well as the reports from the numerous Portuguese voyages of exploration done earlier (p.25).

"It was their opinion- fully borne out by subsequent events, that Columbus was "*overly optimistic*" about his estimate for the distance to Asia."

They refused to fund his expedition on this basis!

What is remarkable is that, they refused not because they thought the world was "flat," as is commonly thought, but because they had "*Better Information*" about the "True Size of the Spherical Earth!" and they knew Columbus was wrong!

Columbus next tried Spain and the court of King Ferdinand and Queen Isabella, where in spite of many delays, he finally received "Royal Funding" for his expedition (p.26). Columbus was wrong not only about the "distance to Asia," but also about there being a land mass between Europe and Asia. Had he been right about the second point, he would have had a vast ocean to traverse and would have certainly run out of provisions. Luckily for him, and for history, he was wrong on both counts! Columbus's voyages literally changed the map of the world forever! (p.26)

Using a branch of modern mathematics called "Topology" one can prove that *no* map is without its faults in trying to accurately convey the true shape of the earth. Beyond the problem of unreliable geographic data was the more fundamental problem of how to transfer sets of latitude and longitude data onto a map without producing "gross distortions." Since navigation lay at the heart of economic development, as well as warfare, it was crucial to get this question right (p.29). There are two qualities in a map that navigator's prize: 1) That the "North" points in the "upward direction" and 2) that all compass directions are correct relative to the northerly direction. Any map that has these two qualities is called a "Navigator's Map." The first actual map designed using these principles was made by the Flemish cartographer Gerhard Kremer, whose last name means "merchant" and who went by the Latin equivalent "*Mercator.*" Today the name is all but synonymous with the map projection of the same. In 1541 Mercator created a map of the globe for Charles V, the Holy Roman Emperor, who then commissioned a set of "Surveyor's Instruments."

"The combination of accuracy and beauty seems to have brought Mercator many commissions and considerable wealth and fame, but his primary interest seems to have been map making." He was universally recognized as Europe's "Premiere Cartographer." The map he drew in 1569 is his most famous (p.31). Mercator explained the principles on which his map was made as follows: (p.33)

"*In making this representation of the world we have had to employ a new arrangement of the meridian with reference to the parallels... We have progressively increased the degrees of latitude towards each pole in proportion to the lengthening of the parallels with reference to the equator.*"

In order to have the map come out right, Mercator stretched the map in the vertical direction and the amount of vertical stretching is the same as the amount of horizontal stretching (p.33). Edward Wright gave an explicit "formula," for the amount of "stretching," at any latitude. With his "tables," one could in principle construct a map without understanding the underlying principles (p.33). It was not until 1668, ninety-nine years after Mercator first conceived of the idea for his map, that mathematicians conceived of the idea of applying the newly invented subject of Integral Calculus" to the problem of solving for the exact equations for Mercator's map (p.34). The biggest drawback of his map is the way it progressively distorts the lines of latitude toward the poles.

It was the search for a way to correct for these distortions, which led to many of the mathematical developments in the area of non-Euclidean geometry that would follow. In many ways, it was due to the simple fact that the lines of longitude didn't fit on the map the way they did on the earth, and were not parallel at all, that led to the necessity of developing a "new" geometry based on "new assumptions."

Carl Friedrich Gauss lived about the same time as the composer Beethoven (though it is doubtful they ever met!). Gauss devoted many years to the study of physics, and to the study of Electricity and Magnetism, however, it would be Astronomy where he would make his first contribution. "Ceres" was the first "Asteroid" (meaning "little star") ever discovered.

Though the Italian Astronomer Giuseppe Piazzi had first observed it, it was Gauss who did the orbital calculations to determine where Ceres would "reappear." His prediction was very accurate and "the object was recovered" and many more observations made (p.43).

Gauss is also remembered for his famous "Bell-Curve" also known as the "Error Curve" or "Gaussian Distribution" which describes the Probabilities of random variables in statistics. He was also a "Surveyor" and agreed to lead a land survey of the area around his home in Göttingen, Germany. Though the work was often tedious and time consuming, he once again used a seemingly mindless task as a springboard to a brand new concept, that would have profound consequences. In order to make the earth fit onto a map, he would need to develop a more advanced branch of mathematics called "Spherical Geometry," but again the formulas were only accurate to within a degree or so. Gauss developed even newer and more powerful formulas, which could be applied to any surface, regardless of its shape! (p.52). He introduced the concept into mathematics of "Curvature," an idea, which could be applied to surfaces, which were either "positively" or "negatively" curved.

A fundamental fact about triangles in a plane, is that the sum of their interior angles is 180 degrees. For "Triangles on a Sphere," however, the sum is always greater than 180 degrees! And furthermore depend on the size of the triangle! For small triangles the sum of the angles is only slightly greater than 180 degrees, but for large triangles of the sort that mapmakers were drawing on the earth the sum could be three 90-degree angles! (p.55). One of the consequences of this was another proof of "Gauss's Theorem" that

"*A perfectly scaled map of any portion of the earth's surface is impossible!*"

Gauss laid out his ideas about geometry in a paper in 1827.

In 1829, a young Russian mathematician named Nikolai Ivanovich Lobachevsky, wrote a paper in which he presented an alternative to Euclidean geometry. He didn't call his new geometry "non-Euclidean Geometry," but "Imaginary Geometry"(p.63-4). Many of the original statements of Euclidean geometry remained intact: Base angles of isosceles triangles remained equal, the largest side of the triangle remains opposite to the largest angle and so on. However, some of the most familiar theorems from Euclidean Geometry are no longer true- Like the Pythagorean theorem! In Lobachevsky's geometry, the angles in the triangle depend on the size of the triangle, in all cases, the sum is less than 180 degrees (p.63).

Lobachevsky's work met with little enthusiasm, because it was called "Imaginary" and it appeared in a fairly obscure journa,l and in Russian. But what reaction there was, first among Russians, and later among mathematicians from other countries was all bad! (p.64)

Even more devastating, was the experience of a young Hungarian mathematician named *Janos Bolyai* , who had independently co-discovered "non-Euclidean Geometry" (p.64). Bolyai's father Wolfgang was a life long friend of Gauss, and sent him a copy of his son's work.

A word of praise for the young Bolyai would have set him on a brilliant career, but instead Gauss reviled the work, and claimed to have already done the same thing!

The reason was that Gauss, had in fact, anticipated much of the work done by Lobachevsky and Bolyai, however, <u>he never published his work!</u> (p.65). He did this not because the work was not up to his high standards of achievement, but because he feared the negative reaction that would surely have followed it!

This is one of the crucial moments in both the book and in the history of mathematics!

<u>No fewer than three mathematicians are credited with the co-discovery of "non-Euclidean Geometry!"</u> And this is really the moment that the book leads us to, because the author wanted to show how the ideas from earlier times, led to this predicament. For the simple reason that mankind needed to navigate, and to do this it needed maps, and since no map was an accurate description of the spherical earth, new methods had to come into being. Thus, the birth of a new branch of mathematics: "non-Euclidean Geometry." This new geometry began with a number of new statements concerning lines. The "Parallel Theorem" not only restated Euclid, it contradicted the old Euclidean axiom altogether (p.65). Theorems, which did not depend on the parallel theorem, would remain, those that did would be replaced by new ones (p.66).

As for the value of this new geometry, Lobachevsky knew that it posed a fundamental question of mathematical science. Did the world in which we live look *Euclidean* or was it more like the *non-Euclidean geometry* that was now becoming "overcrowded" with co-discoverers?

Georg Friedrich Bernhard Riemann was born in 1826, about fifty years after Gauss. He was a modest man, who left only a "small volume of work," but its "high quality" is of special importance in mathematics. Riemann's few publications contrast sharply with the outpourings of <u>Leonhard Euler</u>. Riemann's lifetime contributions are collected in only one small paperback volume (p.78). Yet he was a *pivotal figure* in the development of mathematics.

Gauss's reluctance to praise Bolyai was not evident when he was presented with Riemann's doctoral thesis. Instead he embraced it, calling it "Creative, Active, Gloriously Fertile and Original."(p.78).

The essence of his "new conception," was that we should "probe the space around us" by performing "measurements" as Gauss had done for the magnetic field (p.79). He proposed a novel "thought experiment" for measuring the circumference of a circle (p.81). In a Euclidean universe the circumference is 2 times pi times the radius ($2\pi r$). In Riemann's view we have no way to know beforehand whether space is Euclidean or not! So we really don't know what the circumference is! Both Gauss and lobachevsky had expressed this point earlier. Lobachevsky had worked out all of the details of the non-Euclidean solid geometry, including the circumference of a circle. In Hyperbolic geometry the circumference would be greater than 2 times pi times the radius, by a certain amount called the "Space Curvature"(p.82). Zero curvature yields a circumference of $2\pi r$. Negative curvature yields a circumference "greater than" $2\pi r$. Positive curvature gives a circumference "less than" $2\pi r$(p.83). What curvature "measures" is the degree and kind of deviation from the Euclidean model (p.84).

Our experience tells us that Euclidean geometry holds well, as a description of space, on a small scale, but we have no reason to believe that it also holds in the larger intergalactic space. It is in attempting to make this connection, that we are again left groping with the concept of a "Flat" universe, exactly as an earlier age had done in holding to a flat earth! (p.85).

Riemann not only invented the idea of "Curved Space," and explained how to compute its curvature, he proposed a "radically different model for the universe"(p.85). Specifically, he proposed a description of the universe if it happened to be in the shape of a "Spherical Space" (one of constant positive curvature)(p.85). Such a depiction of the universe would only be artificial because we know that all maps are really distortions of reality (p.87).

He developed a "Thought Experiment"(p.80, 83, &88) that went like this:

1. He envisioned a Bright "halo" of light encircling the planet at a given distance above the earth.

2. To carry out the "thought experiment" let us suppose that we can measure the circumference of that circle (If space were Euclidean then it would be "2 times pi times the radius," but in Riemann's view this "could not be assumed!," because we have no way of knowing it in advance!)

3. We can draw maps where the distances from the center are drawn accurately, but then the lengths of concentric circles become more and more distorted the farther away you get from center, since the positive curvature of the earth's surface results in circles that are smaller than the circles indicated by the corresponding circles on our flat map.
On a flat map those circles just keep getting bigger, while on the earth they reach a final maximum length- A "Great Circle." Then they contract back to the "antipodal point."
If space had a fixed positive curvature in a Riemann sense, the "circles of light " in this experiment would get longer and longer, though not as quickly as in a flat Euclidean space (p.88) and they would eventually reach a maximum limit in our part of the universe. That would be the part of the universe farthest away from us.

4. One feature of this model of the universe that pleased Riemann was that it solved the "age old paradox" about the "Edge of the Universe." Philosophers had speculated about an infinite universe that went on forever. This idea had been rejected by a number of prominent thinkers including Newton. The argument went something like this:
"If the universe didn't go on forever, then had to end somewhere and what was on the otherside of that?"

5. Riemann's model solved this paradox, which is rooted in the belief that the universe is flat, or Euclidean. If instead it has "positive curvature" (a Riemannian surface), then it can be "Finite" in extent and still not have an "edge."
6. In Riemann's model every part of the universe looks like every other part, as far as shape and measurements go (p.88).

Albert Einstein had this to say: "*Only the genius of Riemann, solitary and uncomprehended, had already won it's way, by the middle of the last century, to a new conception of space, in which space was deprived of it's rigidity, and in which it's power to take part in physical events was recognized as possible*"(p.79).

Riemann's conception of a spherical space, together with his suggestion that such a space could describe the actual shape of our universe, constitute one of the most original and radical departures from the standard world view in the history of science (p.91).
Einstein would incorporate Riemann's spherical space into his own work, along with two other important ideas from Riemann 1. the curvature of space and 2. the description of a four dimensional curved space.
Riemann had invented all of these concepts while still in his twenties. As well as "hyperbolic geometry." He presented them at Gottingen in 1854 in a lecture, now seen by many as the birth of Modern Cosmology."
Another great mind in the form of Albert Eistein would come along a half century later to complete the picture of the "Curved Four Dimensional Universe." While still in his twenties Riemann had set himself to the work of establishing a "Unified Mathematical Theory" connecting Electricity, Magnetism, Light and Gravity. This work was so far ahead of its time that a century later Einstein would work, in vain, to create a "Unified Field Theory" in the latter years of his life. To this day physicists are trying to work out its principles, but as in Riemann's time, it's a daunting task. (p.92).

This is the second major point of the book, that Riemann's work anticipates that of Einstein by about fifty years. He developed the concept of "curved space," and even advanced the idea of forming a coherent "unified field theory." You might even be tempted to suggest, that if Riemann had know about the Michelson- Morley experiment, he might even have beaten Einstein to the special theory of relativity! But, this didn't happen, largely because developments in another area of science were yet to be discovered, described and explained.

Next the book talks about the development of "Electromagnetism," the discovery of the "Electromagnetic Spectrum" and James Clerk Maxwell, who in many ways realized a portion of Riemann's quest for a "unified theory" in the area of Electricity, Magnetism and Light (p.98). The year 1800 saw the discovery of "Infrared Light" (William Hershel) and in 1802 ; "Ultraviolet Light" (William Ritter).
The other major scientific discovery of 1800 was the "Voltaic Pile" of Allesandro Volta.
Electric lights would not come until the end of the century (p. 94). The year 1885 saw the discovery of both "Radio Waves" (Heinrich Hertz) and "X-rays" (Wilhelm Roentgen)(p.95).
It was not until later that all these different forms of rays were identified as all being different forms of "Electromagnetic Radiation."
The first Nobel Prize was awarded in 1901 to Wilhelm Roontgen for the discovery of X-rays, which had already shown their future value as a tool of medical diagnosis.
What is surprising is that these discoveries were made mathematically before they were made experimentally (p.95). Much of the credit for this goes to Maxwell. His most famous achievement was the discovery of the set of equations, which bear his name (p.97). Maxwell's equations had an unparalleled ability to predict new phenomena involving electricity and magnetism.
It seemed to be a cornucopia of new inventions (p.99). Their development had been a by-product of his personal search for a "Unified Theory of "Electricity, Magnetism and Light." Maxwell had inadvertently discovered a portion of Riemann's "Unified Field Theory"(p.98).
"Light" was itself a form of "electromagnetic radiation." Maxwell concluded that there should be "other forms" of electromagnetic radiation beyond visible light and soon scientists were finding "Microwaves" and "Gamma Radiation" (p.99). Another surprising consequence of this discovery was that this radiation had been bombarding the earth throughout its history! (p.99).
After WW2 a man named Grote Reber wrote a paper entitled "*Cosmic Static*" which would gave birth to "Radio Astronomy" (p.99). He had built a homemade radio telescope in his backyard in Wheaton, Ohio and had produced a "Contour Map" of a portion of the sky with it based on the strength of the "Radio Static" (p.99).

 From here, the book goes on to point out that much of the astronomy of the second half of the 20th century has been advanced by creating "new telescopes" for seeing the universe in new and different ways (p.99) and that each of these gives us a more comprehensive view of the universe.

Osserman then turns to the "Observable Universe" and the measurement of "Stellar Parallax."
Friedrich Wihelm Bessel was the first scientist to measure the apparent drift in the position of a star against the background of distant stars (p.102). This required two critical insights: First, since the earth's orbit had a diameter of 2 AU (Astronomical Units ~ 93 Million miles, The distance between the earth and the sun), this distance could be used as a baseline for using simple triangles to measure the displacement in a stars position and thereby it's distance (p.102). The second, was that because light traveled at a finite speed, when we look out into space we are also looking back into time itself! The further away a star is the further back in time it is (p.103).

The discovery of the "Expanding Universe" is often credited to Edwin Hubble, who in 1929 presented the observational evidence that distant galaxies seem to be receding at speeds proportional to their distances (p.104). But, the first hints of an expanding universe were contained in theoretical papers of Albert Einstein and the Dutch Astronomer Willem De Sitter.
Both papers were based on Einstein's theory of relativity, which had been written just two years

before in 1915. Einstein determined that the universe could not be "static," while De Sitter determined that it ought to be "*Expanding*" (or put another way, that galaxies should be receding away from each other).

Hubble acknowledged <u>William De Sitter's</u> role in a letter;
"The possibility of a velocity –distance relationship has been in the air for years- You, I believe, were the first to mention it!"(p.105)

Between 1917 and 1927 there were a number of theoretical and experimental advances, which helped to solidify the case for an "expanding universe."

In 1922 <u>Vesto M. Slipher</u> prepared a list of forty-one galaxie,s along with their recessional velocities. They all appeared to be moving away from our own.

In that same year a Russian Mathematician named <u>Alexander Friedmann,</u> found a solution to Einstein's equations, which did away with the need for a constant to prevent the universe from "expanding" and so created a new view of cosmology where the universe could go through "*cycles of expansion and contraction* (p.105)."

<u>Hermann Wyl,</u> the leading German mathematician of his day, had created a model wherein the "galaxies" were "*receding at a rate proportional to their distance.*"

What was missing in these discussions was *a reliable way to measure the distance to far off galaxies.* The solution to the problem came from <u>Edwin Hubble's</u> discovery of a new kind of star in the Andromeda galaxy called a "*Cepheid Variable*" in 1923. Hubble's discovery laid a foundation for settling the debate over the distance to galaxies and also the more fundamental question, "What were the building blocks of the universe made of?"(p.106). Hubble was able to place the "Andromeda Nebulae" as it was then called, well beyond our own galaxy. Between 1923 and 1929, Hubble went on to establish the distances to another 23 galaxies and estimated the likely distances for 21 more. During the same period two cosmologists: <u>Georges Lemaitre</u>, in Belgium and <u>Howard Robertson</u> in the U.S. published papers based on Einstein's equations which showed that distant galaxies would be receding at a rate proportional to their distances (p.107). Once Hubble had obtained a set of reliable distances to a number of galaxies, he was able to compare them with velocities found earlier by Slipher and others and thus the velocity- distance relation known as "Hubble's Law" was established.

Chapter 8 is devoted to Albert Einstein and beyond the mere "presence" of Einstein as a scientist it also describes the three seminal papers he wrote in 1905. The first, "*The Special Theory of Relativity*" recast the basic way in which we think about both the microscopic world where things like atoms are very small and move very fast and the macroscopic universe where galaxies are very large and move very fast (p.128). The Second paper was "*The Photoelectric Effect*" wherein he describes how "electricity" can be derived from "light" and the material properties of certain atoms (p.129). This paper became the foundation for the "Quantum Theory." The third, "*The General Theory of Relativity*" is the most complex and the farthest reaching in its consequences, especially with regard to gravity, curved 4-space and matter (p.132).

Osserman goes on to describe Einstein as the "pacifist opposed to Hitler," also as an "iconoclast," and someone who rarely followed "conventional wisdom." Finally, for his "good natured" bravado and wry aphorism, which are still quoted both in and out of context by everyone, to this day (p.127). Moreover, there are other aspects of Einstein's life, which though not in the book, but are still very important, like his work, as a clerk in the "patent office." This phase of his life is usually interpreted that he was in some way "underemployed," but I don't believe so. If anything really "useful" had come from his theories, then the discussions about how to "capitalize" on those ideas, would have certainly taken place in a patent office. Einstein was a very "astute" and "practical" individual, who realized the importance of "Intellectual Property." He was accorded a PhD subsequent to earning the Nobel Prize in Physics, but his college degree was an "AB in Science" and not even a BS in physics, like we all now earn. Some people even suggest on the basis of his transcripts in Italy, that it was "Electrical Engineering" that was his passion. His father had owned an "Electric Factory" ("Ein Electrishe Gesellshaft" in German). It is ironic that an "Electrical Engineering student" should be held up as the very "model" for all subsequent physics students to follow.

Everything about the life of Einstein has been studied and interpreted and its value distilled into the practical fabric of the college physics curriculum. The courses that he took, we take. The insights that he acquired, are the same ones that we are guided to acquire. He casts a long shadow over all of the physicists of the 20th and now the 21st centuries.

I suppose that it goes without saying that at times, he was very "*out of place in the world.*" Training physicists to be "*deliberate iconoclasts*" for today's business world, with it's own very competitive, and deliberate demands for "conformity" and "submission" of scientists to government and business leaders, is very anachronistic. "*Free Thinkers,*" like Einstein, are as truly out of place in today's business world as they were in their own. After a century of "over development," physics is sadly no longer the central science it once was. College graduates with skills in "Energy and Matter" are hardly noticed anymore against the much larger population of "Information & Computer Science" graduates. Everything to do with "physics" has been "standardized." In essence, we have done all of the physics that will ever be "commercially viable," and a business world, unable to profit from the increasingly "obscure physics" that we "*are*" doing, is letting us know it!

If Einstein were alive today would he still study physics? I personally think not! there's just no "elbow room" in it for scientists anymore! I personally think he would have been some kind of computer scientist in today's world and skipped the physics!

Osserman discusses the properties of Space-Time and Abstract geometry (soap bubbles and chaos theory). The book ends by discussing the concept of "manifolds" and with another discourse on "Chaos."(p.161-3).

And there you have it! The book takes 170 pages to tell this story about "Modern Geometry" and it's influence on the "Mapping of the Universe." The book makes a few sidebars to discuss the results of the COBE mission ("Cosmic Background Explorer"), which mapped the "Microwave Background Radiation" from the Big Bang using very sensitive radiometers and found that the temperature variation in the background is about +/- 6 Nano- Kelvin Degrees. A very smooth and very small result for the microwave radiation distribution throughout the whole observable universe! Osserman views this as a "new and more comprehensive view of the universe." In fairness, it should be pointed out that this mission had already been done from the ground for many years with radio telescopes and we already knew the result! Ten years ago at Mt. Palomar when I was working for Caltech, I remember sitting in on a "bull session" with the scientists about this particular mission before if ever flew and we discussed the +/- 6 kelvin degree variation! That was way back in 1989!

I thought the book did a nice, concise, job of telling the whole story of modern geometry and the stories of the lives of the scientists (both the famous and the infamous, the lucky and unlucky ones) who developed it, the various ancient "classic books" which became the very models for modern scholarship and papers they wrote to bring it to it's current level of sophistication. I was especially impressed with how it handled the "transitions between the ancient and more modern ideas," showing how the older ideas influenced how the modern formulation came about. The historical and anecdotal stories were also very interesting. For me, the history really tells you a lot about how the great minds thought about things, about what was important to them as people and why? I think there is a real need to do "Science History" otherwise "new science" just seems to come out of nowhere! If you want to think of it as a type of "Integrated Science" then that's fine, but I think we, in physics, all agree that a whole lot has gone before us and that it's getting much harder to make the kind of "world class" scientific achievements that built the careers of previous generations. In order to advance the work in our field it becomes increasingly desirable to investigate "causes and effects" of previous discoveries. I think that the traditional M.S. in physics is out of step with the "real professional needs" of most graduates. Maybe we need a more "Liberal Arts" based masters degree in something like "Integrated Science" that would allow graduate students another choice of direction besides "pure" research. Like I pointed out, even Albert Einstein had a Bachelors of Arts degree and not a B.S.! Sometimes "new ideas" born of fresh "creative" insights and the people who formulate them just don't fit the mold! There is more to science than a lot of mathematical formulas quoted from textbooks. Some of the most celebrated scientists had to create their own textbooks from whatever they could find on the subject. It was this "struggle" to find answers to concrete problems, which gave rise to the historical developments described in Mr. Osserman's book. I don't know if the "String Theory" will ever produce a "one size fits all" theory of everything, or if it will become "automated" and put all the physicists in the world out of work, but I hope not!

I think as long as we can still ask "good questions" then the work of science will continue and, in time, even something as "obscure" and "intangible" as "String Theory" may prove its value by holding the answer to some future problem (Maybe even a real Grand Unified Theory).

Mechanics: A Short Introduction to Newton's Laws ("Defining and measuring what matters most, when matter is in motion")

Lets start by defining what we mean when we speak of "objects in motion."

To begin, we must explore the ideas put forth by the British mathematician Sir Isaac Newton. In his most well known work, a text entitled "Philosopiae Naturalis Principia Mathematica" (Mathematical principals of motion) he describes three laws:

1. *All bodies which are at rest tend to stay at rest unless they are acted upon by another body or bodies.*
This is the so called "Inertia Law," because all bodies seek to become motionless unless something causes them to move!

2. *The "Inertial Force" on a body is directly proportional to the time change in its momentum, and this force can be expressed as the product of mass times acceleration*

$$F = dp/dt = \Delta p/\Delta t$$
Or
$$F = Ma$$

3. Finally, the last of Newton's laws is often referred to as the "Equal and Opposite Law." And can be stated as "*For every Force there is an equal and opposite Force.*"
Try to think in terms of the balance between your weight Force on the ground and the restraining force the ground provides to hold you up!

$$F_{acting} = -F_{restraining}$$

Along with these three laws of motion, he also produced a formulation for "Universal Gravitation" to describe why the Moon doesn't in fact, "hit the earth!"

$$F = G \, m_1 \, m_2 / R^2$$

Another important concept in physics is that of "Energy Conservation," which for our purposes will always involve the transformation of *"potential energy"* (energy derived from position or height, or condition, such as the stretching of the rubber band), into "*kinetic energy*" (energy due to motion, like when the stretched rubber band is released). The following table should help you to remember what the basics quantities in mechanics are and how they're computed British units are included because they are the ones we will be working with in our experiments:

Quantity (SI units)	British units	Equation	Description
Mass (Kg)	Pounds (lbs)	M= mass	The mass of an object
Velocity (m/ sec)	Feet per second (ft/sec)	V = velocity	Velocity is the rate of travel
Momentum (p) (Newton meters per second)	Foot-pounds per sec (ft-lbs/ sec)	P =mv	Momentum is the impulse or product of mass times motion
Acceleration (a) (meters per second squared) (m/sec ^{2})	Foot per second squared (ft/sec ^{2})	_ a = acceleration	Acceleration is the change in time of velocity.
Force (F)	Pound (Lbs)	F = Ma = dp/dt	Force is the "applied push or pull" on an object.
Work (Joules)	Foot-pounds	W = Fd Where d is the displacement	Work is equivalent to energy and is the "useful capacity to produce work"

Table of physical terms cont'd:

Quantity (SI units)	British Units	Equation	Description
Elongation (meters) Or percent (%)	Inches or Feet Or percent (%)	$E = (X - X_o)/X_o$	Elongation is a dimensionless unit expressed as a percentage of change in length
Kinetic Energy (Joules)	Foot-Lbs (Ft- Lbs)	$K = \frac{1}{2}mv^2$	Kinetic Energy is energy due to motion.
Potential Energy (Joules)	Foot-Lbs (Ft- Lbs)	$V = mgh$, where h is the height, but may also be the elongation.	Potential Energy is energy due to position or condition.
Power (watts)	Horse Power (Hp)	$P = dW/dt = Fv$	Power is the rate at which useful work is delivered
Simple Harmonic Oscillator (SHO): Period (sec)	Period of oscillations (seconds)	Measured directly using a stopwatch.	It is the time taken by an object undergoing oscillations to make one swing.
Angle (radians)	Radians	Observed directly in integral rotations.	Angular displacement from 0 radians
Angular Velocity (radians per sec)	Radians per second	$S = R\emptyset$, where $\emptyset$ is the angular velocity and S is the linear velocity.	Velocity at which an object spins.
Angular Acceleration (radians per sec ^{2})	Radians per second squared	_a = R^a, where ^{a} is the angular acceleration, and a is the linear acceleration.	The time change in the rate of spin in a rotating object.
Moment of Inertia (Kg - m^2)	Foot squared pounds ft^2- Lbs	$L = I^a$, I is the moment of inertia, and L is the angular momentum.	The moment of Inertia is the measure of inertial mass and its resistance to rotation.
Torque (Kg m^2 rads)	Foot squared pounds ft^2- Lbs	$T = I^a$, where T is the Torque, I is the moment of inertia, and ^{a} is the angular acceleration.	Torque is the "twisting force" of a rotating object.
Hooke's Law		$F = -Kd$, where K is called the "spring constant and d is the displacement or elongation.	Hooke's law describes the restoring force on an object in a spring potential.
Fracture ("Catastrophe"- a final momentous tragic event)		Fracture is both the weight and elongation where the rubber band breaks.	Occurs when the rubber band breaks!

Rubber: Structure and Properties (The Primordial Stuff?)
"A very special "sap" coming from Rubber Plants and Rubber Trees"

Rubber is a very special type of "*Sap*" coming from the "epithelial cells" of certain plants found only in rain forests of South America and Africa, where it is still collected in the traditional way by local tribesmen.

The most common commercial type of rubber is called "*Polyisopropene*" and is the material from which the rubber bands found around newspapers is made.

This material is prized for its "*elasticity*," because it is the type of chemical called an "*Elastomer*." The "*Toughness*" of this material is a measure of how well it can resume its original shape after being stretched, a quality making it perfect for applications where a material needs to be soft, light , and able to resist wear such as automobile tires and the soles of running shoes!

Rubber is also the kind of material called a "Polymer" which means "many units" and really describes the many repeated long chain molecules which comprise its volume. In many ways the Rubber band looks very much the same way on the inside as the outside!

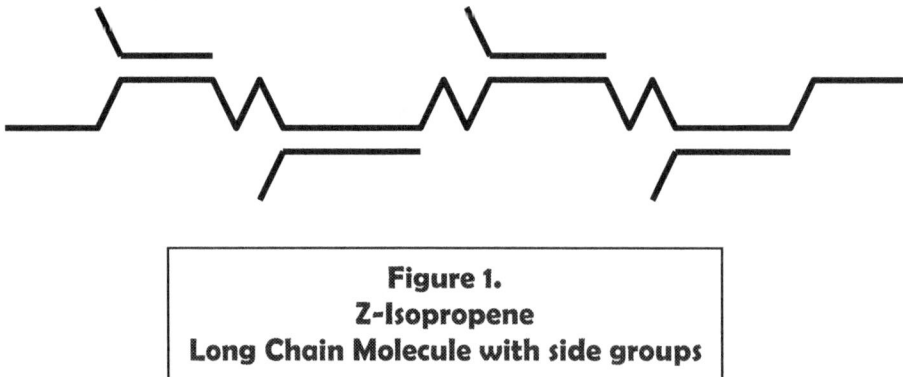

Figure 1.
Z-Isopropene
Long Chain Molecule with side groups

The Greek philosopher "*Democritus*" believed that all matter was composed of very small particles of matter called "*atomos*," which means "*uncutable*" and that all matter was eventually made up of these tiny pieces of uncuttable material. This certainly would seem to be the case for Rubber Bands since no matter how small a piece of the original material you create, you always have a small rubber band that acts just like the whole rubber band!

Commercial Rubber is processed from the natural rubber sap called "*Hevea Rubber*" by a process called "*Vulcanization*," which helps to build both strength and toughness into the material which forms many long hydrocarbon chains from about 1,000 to about 5,000 isopropene units in length. Even after the vulcanization process, in which Sulfur gas is bubbled through the raw Hevea, the processed rubber usually holds only about 1-2% sulfur.

These chains form in randomly coiled bundles, bound by "*Van Der Waals*" forces within the material, produced by side groups in the long molecules to create a "*Chaotic Structure*" within the "*Amorphic*" (meaning "lacking structure") rubber. In truth, the structure isn't amorphic at all, in fact there's so much of it that its complexity looks so irregular that it appears from the outside as if there were no structure when in fact there's so much structure that no single unique structure can be determined. This is an example of "Chaos in nature," where a simple unit is repeated so many times that the individual variations overlap so frequently they create a pattern so complex that we simply call it "*Chaos*." If you have ever played with home computer programs for making "fractals" you might know what I mean.

When Force is released from the rubber band, the polymer chains do not return completely to their original positions, and instead remained displaced in a process called "Deformation." Because the intermolecular forces are only very weak, an external deforming force can not only stretch the coiled polymer but can allow it to split by breaking mono- and di-sulfide cross-links which form between the chain molecules to provide the material with additional rigidity, but if too much cross-linking takes place, then the result is a rigid rubber without its usual elastic properties.

"*Toughness,*" the ability of the material to regain its shape after stress and strain have been applied to it, is also a measure of the energy needed to deform or even fracture the material. It is the same energy as would be found under the "stress-strain curve."

Rubber is an electrical insulator with an energy gap greater than 4eV. In the case of polyisopropene that gap is about 13eV! Insulators still display conductivity as given by the familiar relation: conductivity in mhos is given by the product of $nq\mu$, where n is the number of charge carriers in the conductor, q is the charge of one electron, and μ is the "electron mobility" of the carriers through the material. The "drift velocity" of the charge carriers can be determined by the product of mobility and E- field: $V = \mu E$

Chemical Theromodynamics and Electron Gases

Point Mass Gas and Particle Velocities:

Also, don't forget that the rubber band is also like a volume of gas with "point masses" and that its thermal energy is E = 2/3 U, where U is the energy of an "ideal gas" (U = 3/2 KT), and therefore the total thermal energy of the rubber band is E = KT , where K is Boltzmann's constant (K = 1.38 x 10 -23 Joules / °K) and T ~ 320 °K. Setting this value equal to the kinetic energy allows us to solve for the "*average velocity*" of the "*point mass gas particle*"

(3.1) $V = (2KT/ m)^{1/2} = (2(1.38 \times 10^{-23} K/ °K)(300 °K)/ (1.99 \times 10^{-23} Kg/ Carbon\ atom))$
= 20 m/s (Or about 66 ft/sec)

We can compare this result to the "Maxwellian average velocity:" $V = (KT/m)^{1/2}$ = 14 m/s
And to the "Most Probable Veleocity:" $V = (3KT/m)^{1/2}$ = 25 m/s
To discover that the Maxwellian average velocity is the lowest, the average velocity is in the middle and that the "most probable velocity" is the highest velocity of the gas particles!

(3.1A) (Maxwellian < Average < Most probable)
$(KT/ m)^{1/2} < (2KT/ m)^{1/2} < (3KT/ m)^{1/2}$
(14 m/s)   (20 m/s)   (25 m/s)

Energy of the Conduction electrons:

An "Electron Volt" is a "CGS"-unit for electron energy, and represents the amount of energy imparted to an electron at rest from being accelerated through a potential of 1 volt in 1 centimeter. We use this unit when we work with its chemistry.
We can also express the Boltzmann's constant in terms of Electron volts:
$K = 0.86174 \times 10^{-4}$ eV/ °K

In doing this we can now express the *total energy of our "point mass gas"* as

(3.2) $E = KT = (0.86174 \times 10^{-4}$ eV/ °K) (300 °K) = 0.259 eV
This is equivalent to the energy of one electron accelerated by a 259 milli-volt potential for 1 cm (a very low potential energy!

The *"Thermal Wavelength"* of our rubber band can be determined from the following expression:

(3.3) Thermal wavelength = $(h^2 / 2\pi\ m\ KT)^{1/2}$
= { (6.625 x 10 $^{-34}$ J*s) / 2π (1.99 x 10 $^{-23}$ Kg/C)(1.380658 x 10 $^{-23}$ J/°K) (300 °K)
= 9.2 x 10 $^{-13}$ = 0.92 pm (pico meters) (high energy gamma rays)
Where h is Plank's constant (h = 6.625 x 10 $^{-34}$ J*s) and m is the mass of the ideal point mass particles within the gas (~ 1.99 x 10 $^{-23}$ kg/ Carbon atom).

29

Solid State Plasma Effects in Insulators:

Our rubber band's energy gap creates a region which contains a "solid state plasma" with many properties in common with other naturally occurring space plasmas like the solar wind or even the Auroras (both Borealis & Australis), and there's still a lot of "empty space in the rubber band!

One way that it does this is by containing gases, electrons, photons, and a number of "quantized vibrations" called "phonons" Because it behaves as an electrical insulator, it has a large "energy gap"(~13eV) which tends to discourage all but the most powerful cosmic rays from causing conduction within it. The diagram below shows the essential processes in the energy gap.

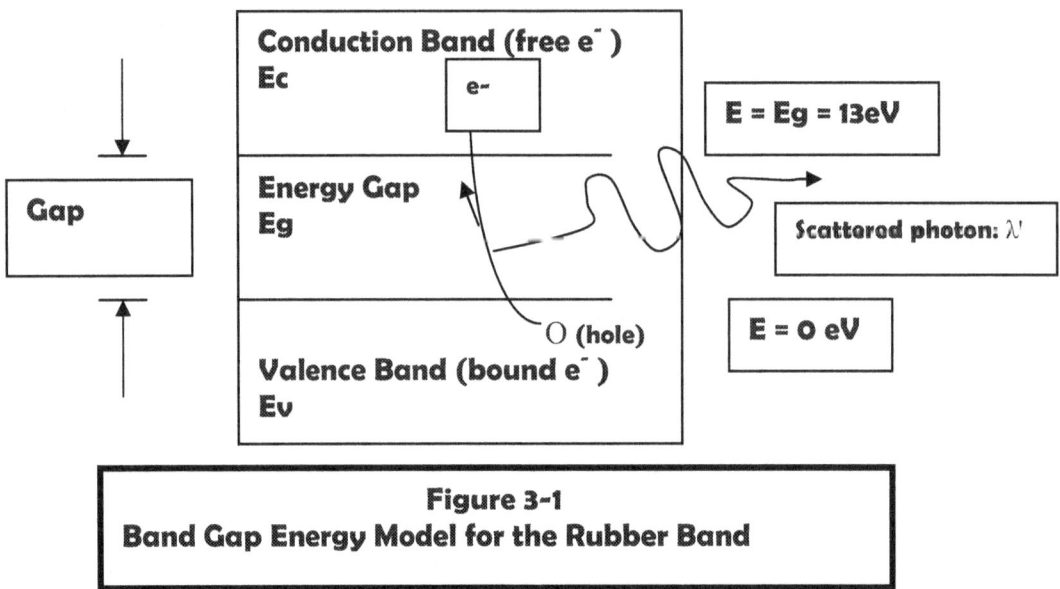

Figure 3-1
Band Gap Energy Model for the Rubber Band

An "Exciton" is the average separation between an electron/ hole pair created in the area of the gap, by the conduction of free electrons and their depletion from bound sites.

There are seven basic processes in high energy physics: 1. pair creation, 2. pair annihilation, 3. Compton scattering, 4. Inverse Compton scattering, 5. Bremstaulung, 6. Synchrotron radiation, and 7. Cherenkov Light. All of these processes occur within the energy gap of the rubber band. "Compton Scattering" occurs when an gamma ray thermal photon collides with a bound electron within the valance band, giving rise to a scattered gamma ray photon with a slightly lower energy and a "recoil electron" which conducts across the gap into the conduction band. The processes are described by the following relations:

(3.4A) $(hc/\lambda) + m_oc^2 = (hc/\lambda') + \gamma m_oc^2$

(3.4B) $(hc/\lambda) = (\gamma m0\beta c2 \cos\varphi + (hc/\lambda') \cos\theta$

(3.4C) $(hc/\lambda') \sin\theta = \gamma m_oc^2 \beta \sin\varphi$

This system can be solved to yield the following result for the wavelength of the scattered photon
(3.4d) $\lambda' = \lambda c (1- \cos\theta)$, where $\lambda c = h/m_oc$ and θ is the scattering angle.

Once the resting electron has been scattered by the incident colliding photon, the conducting recoil electron and the scattered photon with reduced wavelength are either reabsorbed into the material or else they may participate in annihilation processes:

(3.4e) $e^+ + e^- \rightarrow \lambda$ (pair annihilation producing a photon with energy = $2m_oc^2$)

Or

(3.4f) $\lambda \rightarrow e^+ + e^-$ (pair creation producing an electron and a positron from a photon)

The Compton scattering process is shown in fig 3-2

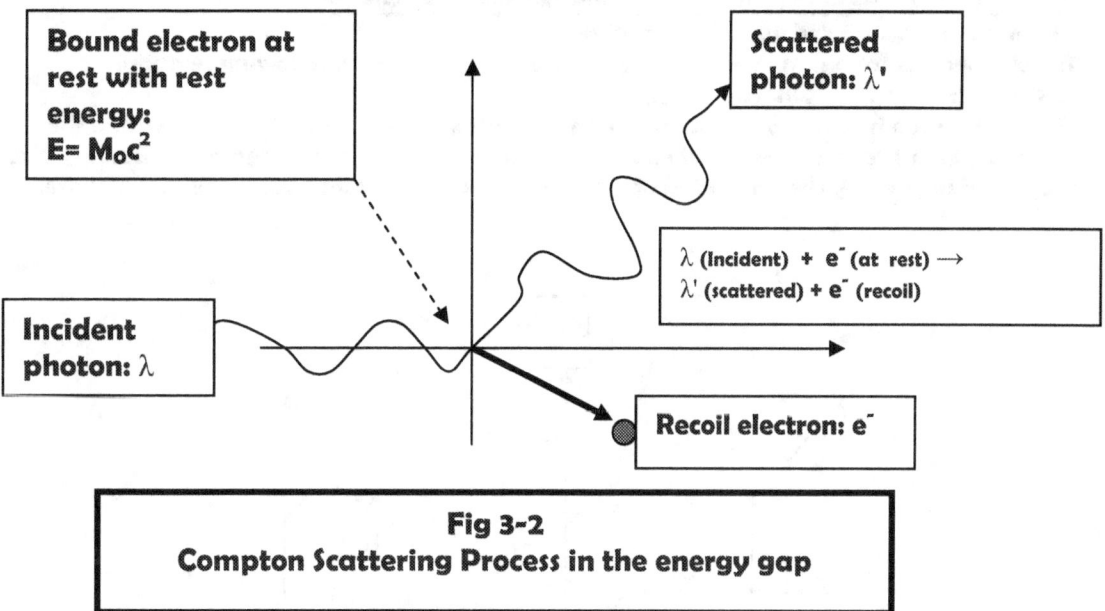

Fig 3-2
Compton Scattering Process in the energy gap

Labels within figure:
- Bound electron at rest with rest energy: $E = M_0 c^2$
- Scattered photon: λ'
- λ (Incident) + e^- (at rest) $\longrightarrow$ λ' (scattered) + e^- (recoil)
- Incident photon: λ
- Recoil electron: e^-

The *number of free electrons* in the energy gap can be determined from the following integral:

(3.4) $n = \int_0^{E_f} g(E)\, dE$

where $g(E) = \{ 8(2)^{1/2} \pi\, m^{3/2}\, E^{1/2} \}/ h^3$ and is called the "density of states"

The solution turns out to be quite easy (just an integral of $E^{1/2}\, dE = 2/3\, E^{3/2}$, the result gives

(3.5) $n_f = (8\pi m^{3/2})/ 3(2)^{1/2} h^3\, E_f^{3/2}$
$= \{8\pi\, (1.99 \times 10^{-26}\ \text{gm/C atom})^{3/2} \}/ \{ (4.24264)(4.136 \times 10^{-15}\ \text{eV*s})^3\, (13\text{eV})^{3/2} \}$
$= (5.0 \times 10^{-25} / 1.41 \times 10^{-41}\ ) = 3.55 \times 10^{16}$ free electrons

The *Fermi Energy* is given by

(3.6) $E_f = (3nh^3 m^{-3/2} / 16(2)^{1/2} \pi)^{2/3}\ = (h^2 / 2m)\, (3n_f / 8\pi)^{2/3} = 7.5\ \text{eV}$

The *number of conduction electrons* in the energy gap is given by

(3.7) $n_c = (8\pi m^{3/2})/ 3(2)^{1/2} h^3\, [E_c^{3/2} - E_f^{3/2}]$
$= \{8\pi\, (1.99 \times 10^{-26}\ \text{gm/C atom})^{3/2} / (4.24264)(4.136 \times 10^{-15})^3\, [(13\ \text{eV})^{3/2} - (7.5)^{3/2})]$
$= (7.06 \times 10^{-38} / 7.90 \times 10^{-42}) = 8.93 \times 10^9$ conduction electrons (fewer than n_f !)

Fewer electrons contribute to conduction than the number of free electrons!

(3.7A) $n_c = nf_0\, \exp(-E_g / 2KT)$

With an energy Gap of 13eV the photon that would be emitted if one could be would be at a wavelength given by
(3.8)   wavelength $= hc / E_g = \{(4.136 \times 10^{-15}\ \text{eV*s})(3.0 \times 10^8\ \text{m/s})\}/ 13\ \text{ev} = 95\text{nm}$ (gamma rays!)

Phonons: Quantized Vibration States

"Phonons" are quantized vibrations in solids.

The eigenvalues for a simple harmonic oscillator are given by the following relation:

(3.8) En = hω (n + 1/2) , where n = 0,1,2,3,...

This relation can be used to graph the" phonon spectra" for the vibration states within the oscillator, in this case a "V"-shaped envelope. Phonons are also considered to be "Bosons, that is they are described by the same kind of statistical functions as other Bosons (like the photon).

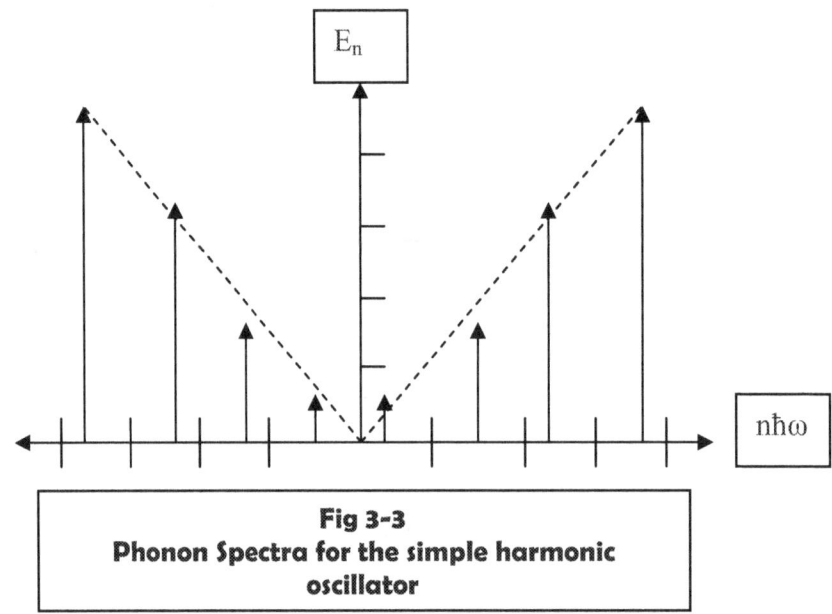

Fig 3-3
Phonon Spectra for the simple harmonic oscillator

The number of phonons in the material can be determined by the following integral

(3.9) $N/2 = \int_0^\infty E_n P_n \, dE$

This integral can be rewritten to

(3.10) $N/2 = (kT)^2 \int_0^\infty h\omega/kT \, (n+1/2) \, (\exp -h\omega/kT - 1)^{-1} \, (h/kT) d\omega$

$$= (kT)^2 (n+1/2) \int_0^\infty (kT)^{-2} (h\omega)(\exp (h\omega/kT - 1)^{-1} \, h \, d\omega$$

This integral is now stated in a standard form

$$\int_0^\infty x \, (\exp (x) - 1) dx = \pi^2/6$$

Now (3.10) can be solved as

(3.11) $N = (\pi^2 k^2 T^2)/3 \, (n + 1/2)$, where n = 0,1,2,3,...

Now we know the distribution function for the phonons. Its shape is the same as shown in figure 3-3 except that it is scaled to show number of phonons on the y-axis.

The Total number of Phonons (**N**) in our "Phonon Gas" must be all phonons in all of the states

(3.12) $\mathbf{N = (kT)^2} \Sigma_0^\infty \pi^2/3 \, (n+1/2)$

Applying the binomial series $n+1/2 = \{ 1/2^{(1)} + (1)^{(0)} + 0 + ... \} = 3/2$

Then $\mathbf{N} = (\pi^2 k^2 T^2)/2 \sim 4 \, k^2 \, T$

32

Creative Experiments using Rubber Bands:
"Now the fun begins!"

Experiment #1: Hookes' Law (weighing things & nuclear potential)
HOOKES' LAW

A spring is a system that obeys Hookes' Law ($F = -kx$). In this spring system there's "Restraining Force" acting against any "displacement from equilibrium" (that's what the minus sign is for. It means a negative (or attractive) force acts inwardly between the ends of the spring as it is stretched).
FORCE (measured in Pounds) = -(K) times (Displacement (measured in Inches)), where K is the "Spring Constant.

QUESTIONS:
Answer the following questions:

1. a) What happens if you take an ordinary rubber band and you apply a force to it? b) What kind of a Force has a minus sign?

2. You are supplied with a set of weights and a stand. Can you think of a way to measure the amount of stretch in the rubber band as increasing amounts of weight are applied? Make a table showing your data for your method (on the next page) and describe how you came to your conclusions (Do a good job! You will need to graph this data so that it can be used again).

Weight (lbs)	Displacement (inches)

HOOKES' LAW cont'd (Graph)

3. a) Display your data in graphical form. b) What do you call the slope of this
    curve? c) What is its value in lb/in?

4. If your family's car weighs 3000 lbs, what length of elongated rubber band would this
    correspond to?

Experiment #1: Hookes' Law Cont'd
Measuring the weight of unknowns

Here you should use the graph of your Hook's Law data to estimate the weights of the various given unknowns. If the Force is proportional to the displacement, how do you think you can measure the force indirectly? How could you improve the accuracy of your method?

OBJECT	DISPLACEMENT (in)	WEIGHT (Force)
Object #1 (Hammer)		
Object #2 (Vice Grips)		
Object #3 (Wrench)		
Object #4		
Object #5		
Object #6		
Object #7		
Object #8		
Object #9		
Object #10		

QUESTIONS:

1. a) What method did you use to make your length measurement and do you think it was very accurate? b) What does Interpolation mean? Explain why.

2. What was the "Average Weight" of the objects measured? (Hint: You can determine this by adding up all of the weights in your table and dividing by the number of objects)

   Average Weight of all the objects: _____ (lbs)

3. Could you weigh even heavier objects (approximately 25lbs let's say) using this method? What might you need?

Experiment #2: Stress/Strain Curve (fracture as a "*catastrophe*")

All materials display both "*Stress*" and "*Strain*" to a greater or lesser amount, it is an inherent property of all matter. Stress is a measure of the strength of a material, while strain is a measure of "*ductility*" - The total plastic strain (stretching before fracture), which is also thought of as the "*effective elongation.*"

"*Toughness*" is the amount of energy required to make a sample fail (i.e. to exhibit "fracture"), in the case of rubber bands, we will apply forces of differing amount to the system, while measuring the "reduction in area" at the cross sectional area of the band.

Stress is the amount of applied force per unit area (like pressure) and has units of Pascals (Pa) or (in our case, because we will be using ordinary yardsticks and fishing weights) PSI (pounds per square inch).

We will be careful to record our data, so that we can display it in a graph from which we will be able to also determine the "*Young's Modulus*" for the sample at any amount of applied force. The "*Tensile Strength*" of the material is determined from the maximum force divided by the cross-sectional area reduction just before fracture.

<u>Parts List:</u>
1. One Stand to hang things from.
2. One Yard Stick
3. One Caliper Micrometer (used to measure the area reduction)
4. Rock Cod Fishing Weights of various sizes
        1lb x 2
        2lb x 2
        3lb
        5lb x 3
5. Rubber Bands (test samples)
6. Data Sheet(s)

<u>Stress-Strain Experiment Set Up:</u>

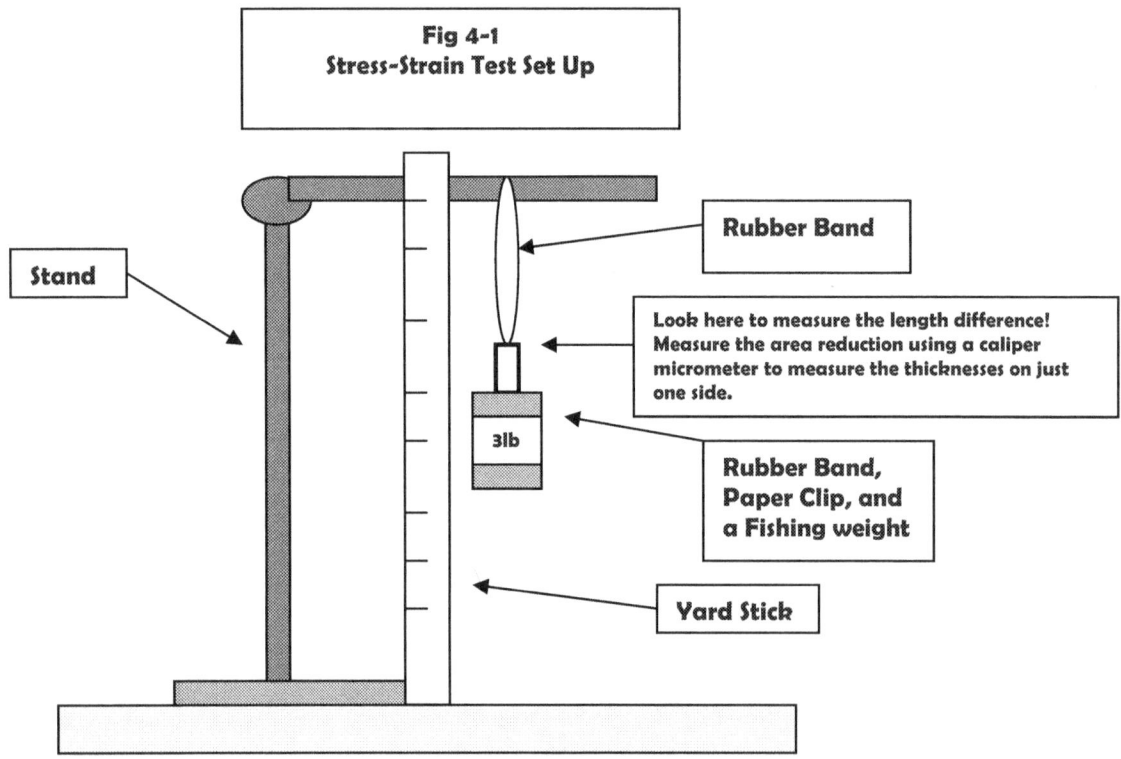

Fig 4-1
Stress-Strain Test Set Up

Rubber Band

Look here to measure the length difference! Measure the area reduction using a caliper micrometer to measure the thicknesses on just one side.

Stand

3lb

Rubber Band, Paper Clip, and a Fishing weight

Yard Stick

"*Deformation*" occurs when forces are applied to a material. "Stress" is the force per unit area, while "*Strain*" is the deformation per unit length. Ductility is the permanent strain induced into the material just prior to fracture and failure.

Many useful design properties depend on an understanding of both "Strength" and "Toughness."

Procedure:

1. Set up the stress-strain test as shown in fig 4-1
2. Measure the initial length of the rubber band and record on the data sheet.
3. Put the rubber band under a 1 Lb load and note the difference in the length of the rubber band. Record this value on the data sheet.
4. Using the caliper micrometer, measure the cross-sectional area of the rubber band in both axial dimensions. Record both of these values in the data sheet.
5. Repeat steps 3 and 4 for all of the available weights (1lb, 2lbs, 3 lbs, etc ...) Note when the rubber band breaks (if it does).
6. Once all of the data has been collected, then the "Stresses" (forces per unit areas) and "elongations" and "strains" (length differences divided by the initial length) can be computed. Also note the errors in the test by noting that the yard stick is not more accurate than +/- 1/16th of an inch. The weights can be assumed to be accurate to about +/- 1% (0.01 lb).
7. When these values have been computed, now we can begin to graph the results. The plot should be arranged so as to put the "Stress" on the y-axis and the "Strain" on the x-axis in order to show the over-all performance characteristics. There may be two sets of linear responses, if so they are the "elastic" and "elastic-plus-plastic" regions of the stress-strain curve. The "elastic" deformations occur when the side groups are squeezed against the molecular axises, while the "elastic-plus-plastic" deformations occur when even the Carbon atoms along the back bone of the molecule have their bonds stretched. The "Plastic" response happens when the rubber band fractures!
8. Different rubber bands can be compared regardless of their physical size and other properties if the Stresses are divided by the "fracture stress" to produce a dimensionless or "normalized stress" so that the graph is completely general, having no units to carry along so all samples may be represented upon it. The "slope(s)" represented on the graph is (are) the "Young's Moduli" for the samples.
9. The Temperature Change due to expansionof the rubber band may be calculated by the following relation:

$$\ln(\sigma_2 / \sigma_1) = R [1/T_2 - 1/T_1], \text{ where } R = 8.316 \text{ Joules/ mol } ^\circ K$$

The idea of *Catastrophism* found its way into the science of Robert Hooke, a member of the very influential scientists of the "Royal Society," in two seminal papers he wrote in 1668 entitled "*Discourse on Earthquakes*," in which he put forth the idea (citing legends) that large movements within the earth itself, were the cause of mountain building. For the geologists of his time it began to seem that the morphology of rock formations could not be explained on the basis of gentle change over long periods of time, but rather showed that very violent forces had acted very quickly, as is the case with volcanos (Bowler, p.37). We now refer to changes that occur "once and forver more" as "Catastrophes." As the rubber band stretches it eventually reaches its limit and breaks. This too is an example of catastrophe.

Data Sheet: Stress Strain experiment

Initial length = X_o = _____   Initial Temperature = _____

Weight (Lbs) (+/- 0.01 lb)	Displacement (inches) (+/- 0.0625")	Elongation E = X − Xo (inches)	W1 (in)	W2 (in)	Area (in²)	Stress (psi)	Strain ε = E/Xo (%)	Temp (°K)
1								
2								
3								
4								
5								
6								
7								
8								
9								
10								
11								
12								
13								
14								
15								
16								
17								
18								
Fracture:								

Data Sheet: Stress Strain experiment (Sample; Sunday Paper)

Initial length = X_o = 6.5   Initial Temperature = 25 °C

Weight (Lbs) (+/- 0.01 lb)	Displacement (inches) (+/- 0.0625")	Elongation E = X − Xo (inches)	W1 (in)	W2 (in)	Area (in²)	Stress (psi)	Strain ϵ = E/Xo (%)	Temp (°K)
1	10.25	3.75	0.048	0.012	0.000576	1736.1	0.577	
2	16.00	9.5	0.013	0.013	0.000637	3189.7	1.46	
3	18.50	12.0	0.050	0.012	0.000600	5000.0	1.85	
4	19.50	13.0	0.050	0.012	0.000600	6666.7	2.00	
5	20.125	13.625	0.045	0.013	0.000585	8547.0	2.10	
6	20.75	14.25	0.043	0.013	0.000559	10,733.5	2.19	
7	21.25	14.75	0.044	0.013	0.000572	12,237.8	2.27	
8	21.50	15.0	0.044	0.013	0.000572	13,986.0	2.31	
9	21.875	15.375	0.041	0.013	0.000533	16,885.6	2.37	
10	22.50	16.0	0.040	0.012	0.000480	20,833.3	2.46	
11	22.875	16.375	0.042	0.011	0.000462	23,809.5	2.52	
12	23.00	16.5	0.041	0.012	0.000492	24,390.2	2.54	
13	23.25	16.75	0.046	0.011	0.000506	25,691.7	2.58	
14	23.50	17.0	0.045	0.011	0.000495	28,282.8	2.62	
15	24.00	17.5	0.045	0.011	0.000495	30,303.0	2.69	
16								
17								
18								
Fracture:								

Experiment #3: Oscillations & Moment of Inertia (& Gyro Effect)
(The following is an exerpt from a unit-plan, done for a Methods course at CSU Fullerton, entitled "Balanced & Unbalanced Forces" by David Tracy)

ANGULAR HARMONIC OSCILLATOR:

At this station you are given a "Rotator" to observe. Set the rotator in motion and describe what you observe.

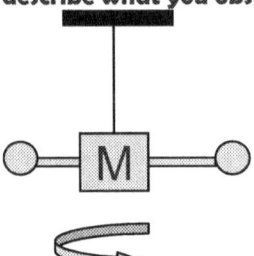

Questions:
1. What are the different parts of the rotator's motion? Describe the "motion" you observe carefully. Do you notice that the motion is more uniform at certain points in the oscillation?
2. How do you explain this?
3. Using a Stopwatch time ten periods of the motion.
    Rotator Period: _____ (sec)
4. Can you build a "Coupled Oscillator" using the rotators? Build this set up and set it in motion. a) Describe what you observe carefully, b) How does adding two rubber bands to the bottom rotator effect the motion?

MOMENT OF INERTIA

Change the mass on the rubber band to either the BLOCK or the SPHERE and observe the motion (I prefer to use 2 different 1 lb Rock Cod Weights for this! You can probably find some at a good tackle shop)

Using a Stopwatch time ten periods of the oscillations and divide by 10 (moving the decimal point to the left one place):

*   BLOCK Period: _____ (sec)
*   SPHERE Period: _____ (sec)

QUESTIONS:
1. a) Are the Period times the same or different? b) Which was the fastest?
    c) Which was slower, d) which was slowest? e) How does the shape of a thing influence its rotation period? Explain your answers.

Experiment #4: Complex motion (Gas molecules & Earthquakes)

You can get a sense of how real motions in nature are from the rotators motion when you set it to rotate, move up and down, as well as sideways. Real motions in atoms, crystals, and molecules would likewise demonstrate such unusual responses. We can model the motions occuring in an earthquake, by using our stress-strain experiment to show what happens to rocks within the earth as forces are built up there. Then, we can use the rotator to show what happens on the surface of the earth when the s-waves (shear forces) and p-waves (up& down forces) hit! There is also usually a lot of shaking (random motions released as heat). The Coupled Oscillator provides an excellent model for "Aftershocks."

A really great "Earthquake and motion detector" can be made using rubber bands and weights as shown in figure 4-1.

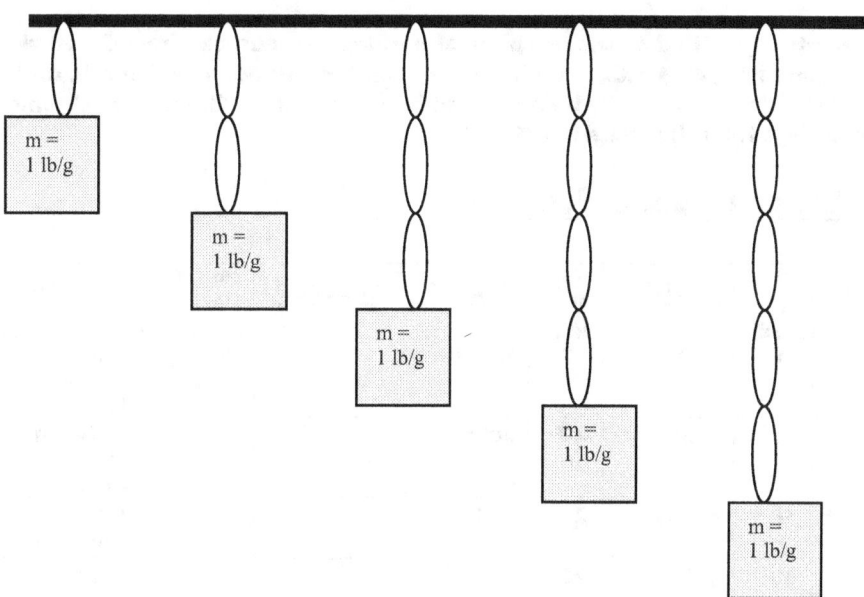

| Fig. 4-1 |
| Model for Earthquake Detector |

The natural frequencies for the various weight/band combinations may be determined from the following table:

Weight	Natural Frequency ($\omega = (1/\sqrt{n})$ $\sqrt{k/2m}$)	Amplitude (inches)
1 lb w/ 1 loop	$(1/\sqrt{1}) \sqrt{k/2m}$	
1 lbs w/ 2 loops	$(1/\sqrt{2}) \sqrt{k/2m}$	
1 lbs w/ 3 loops	$(1/\sqrt{3}) \sqrt{k/2m}$	
1 lbs w/ 4 loops	$(1/\sqrt{4}) \sqrt{k/2m}$	
1 lbs w/ 5 loops	$(1/\sqrt{5}) \sqrt{k/2m}$	

Once a set of vibrations has set the detector in motion, the amplitudes for the various resonance frequencies can be observed, measured, or estimated. (Graph the Phase Space [p,x]).

Experiment #5: Clock Escarpment (It's about time!)

Many problems in physics take the date and time as there starting, or ending points.

Before October 4, 1582 the "Julian Calendar" was in general use, and divided the year into 365 days except for years where the year numbers were divisible by four, when there were 366 days. When we use the "Gregorian Calendar," we are measuring (or trying to measure) the time difference between the present and the exact instant of the "Birth of Christ."

The exact length of the year is the time needed to complete exactly one orbit around the sun, which turns out not be a whole number of revolutions, the length of the tropical year is approximately 365.2422 days (adopting the convention of "leap years" every fourth year).

Pope Gregory VIII ("Gregory the Great") once decreed that the inclusive dates Oct 5, to Oct 14 in 1582 be abolished to correct for the "accumulated error" which had built up in the days on the calendar due to the difference between the "tropical year" and the "Julian year." Gregory's Calendar is still in use today, and was adopted in Britain and the U.S. Colonies in 1752, though the suppression of accumulated errors still continued to be repaired by essentially "throwing away days."

Astronomers eventually settled on the adoption of midday as measured from Greenwich Observatory on January 1, 4713 B.C. as their starting point, or fundamental epoch, and any Gregorian Calendar date can readily be converted into a numerical "Julian Date," which gives an exact number of days from that date in 4713 B.C.

Months of the principal Calendars

Gregorian	No# of Days	Jewish	No# of Days	Muhammadan	No# of Days	Hindu	No# of Days
January	31	Tishri	30	Muharram	30	Chait (March-April)	
February	28	Heshvan	29-30	Safar	29	Baisakh (April-May)	
March	31	Kislev	29-30	Rabi I	30	Jeth (May-June)	
April	30	Tebet	29	Rabi II	29	Asarth (June-July)	
May	31	Shebat	30	Jumada I	30	Sawan (July-August)	
June	30	Adar	29-30	Jumada II	29	Bhadon (August-September)	
July	31	Nisan	30	Rajab	30	Asin (Sept-October)	
August	31	Iyar	29	Sha'ban	29	Kartik (Oct- Nov)	
September	30	Sivan	30	Ramadan	30	Aghan (Nov-Dec)	
October	31	Tammuz	29	Shawwal	29	Pus (Dec-Jan)	
November	30	Ab	30	Dhu'l-Qa'dah	30	Magh (Jan-Feb)	
December	31	Elul	29	Dhu'l-Jiija (in leap years)	29 (30)	Phagun (Feb-Mar)	

Notes:
1. The equinoxes occur on March 21 and September 23, the solstices on June 22 and December 22

Building the Clock Experiment:
A really cool "Clock Escarpment" can be made easily using rubber bands and few wood parts:
Parts List:
1. Gear (made from Cardboard)
2. "T"-piece (made from 3/4" square stock pine, 2 hooks and qty (1) 3/8" Bolt)
3. The "Escarpment" piece (made from 3/4" square stock pine and qty (2) countersunk screws)
4. Gear Shaft (made from 3/4" square stock by 7" in length, with two hooks)

Clock Drawings:

Fig. 5-1
"Gear"
1) Draw two concentric circles 7" and 8"
2) Cut our every other tab to create a gear with eight teeth

Cardboard
"C" = "cut out"

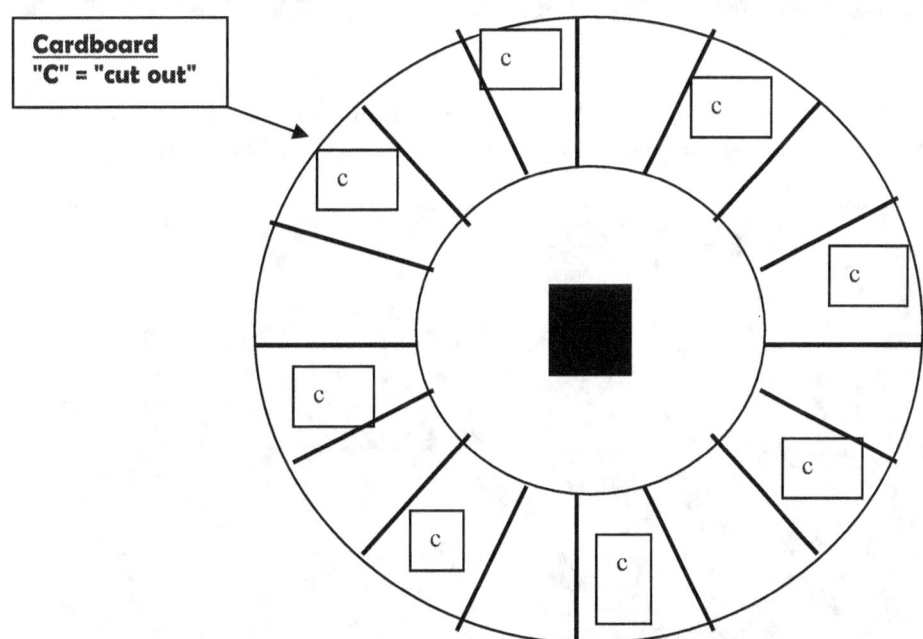

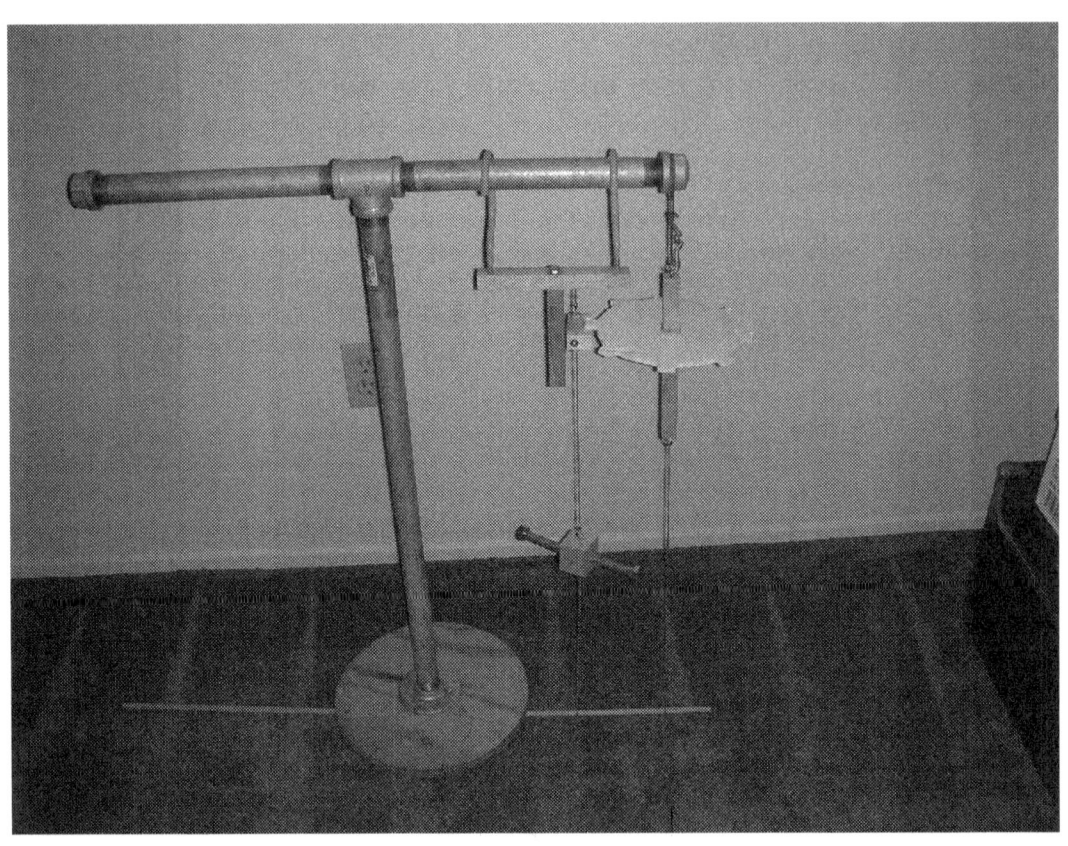

Clock Escarpment: Experiment # 5
It might take a little bit of adjusting to make it work right! (Try using 2 sets of looped bands on the "main gear" (bolt block) to give it both more speed and throw).

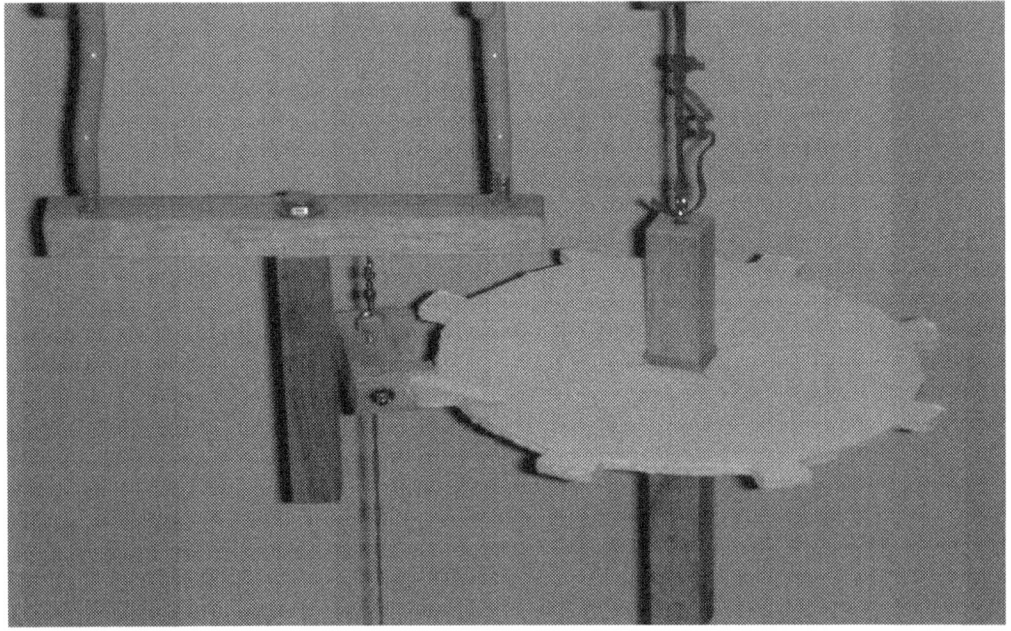

Close Up of the escarpment itself (note the swivel used for the cross-piece).

Electric Generator: Experiment #6
This view shows the coils (150 turns each) and armature assemblies

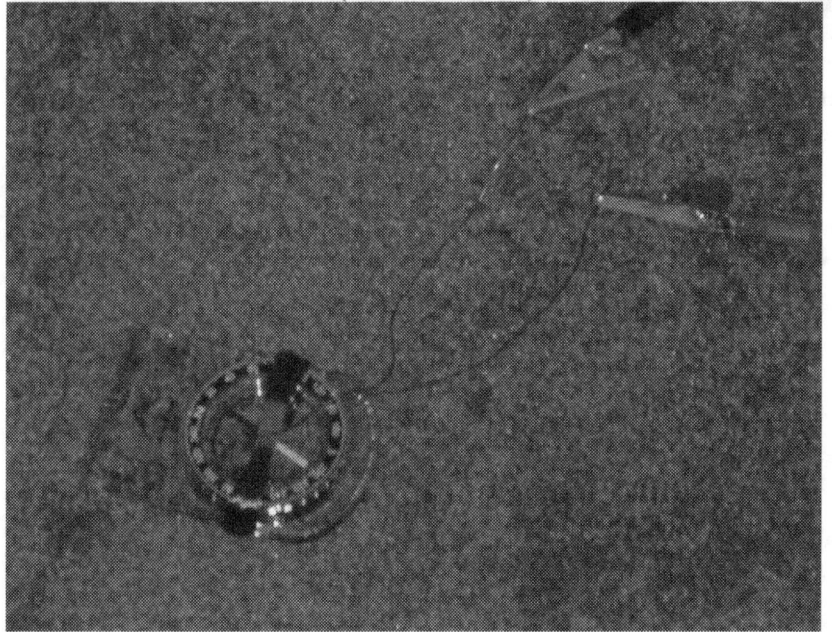

Close Up of the Galvanometer (i.e. a liquid filled compass with about 75 turns of wire and about 6 feet of extra length to put it far away from the influence of the magnets).

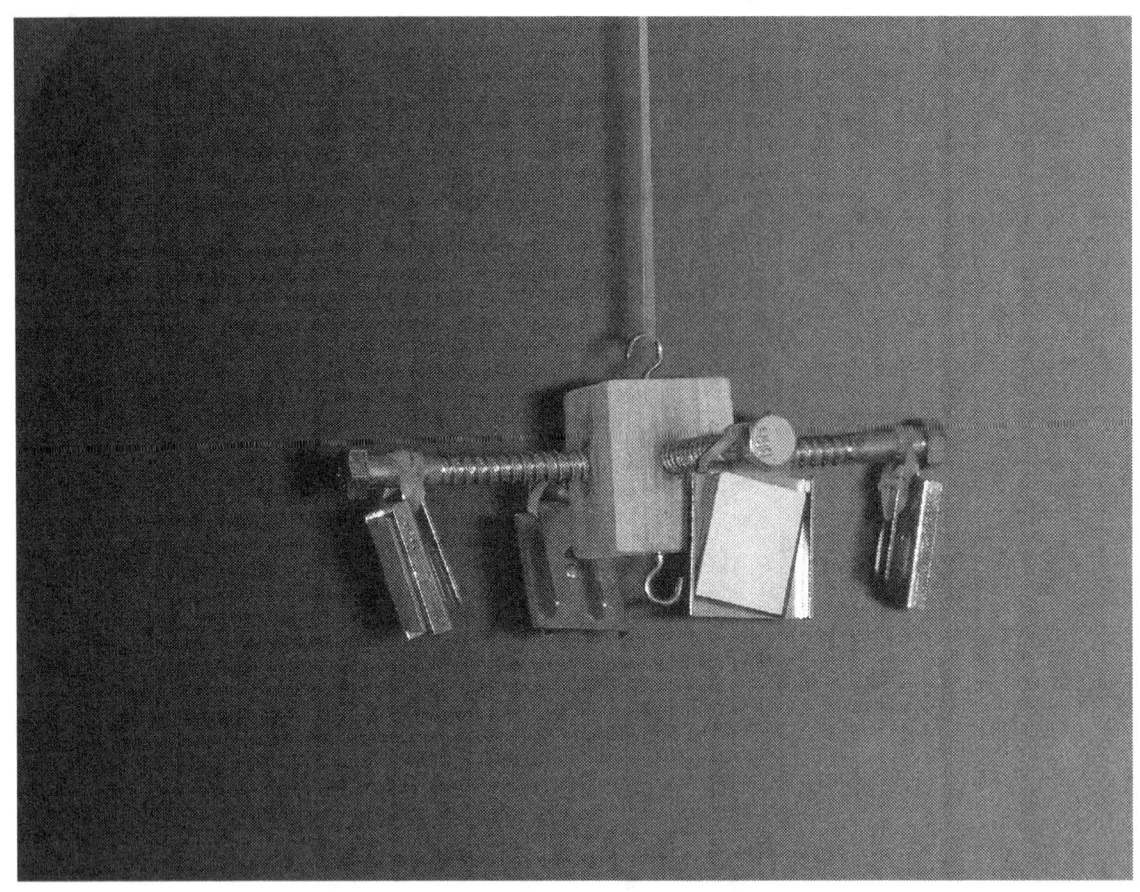

Close Up of the Armature Assembly for the Electric Generator in Experiment # 6, showing the magnets used, as well as how the rubber bands were used to attach them.

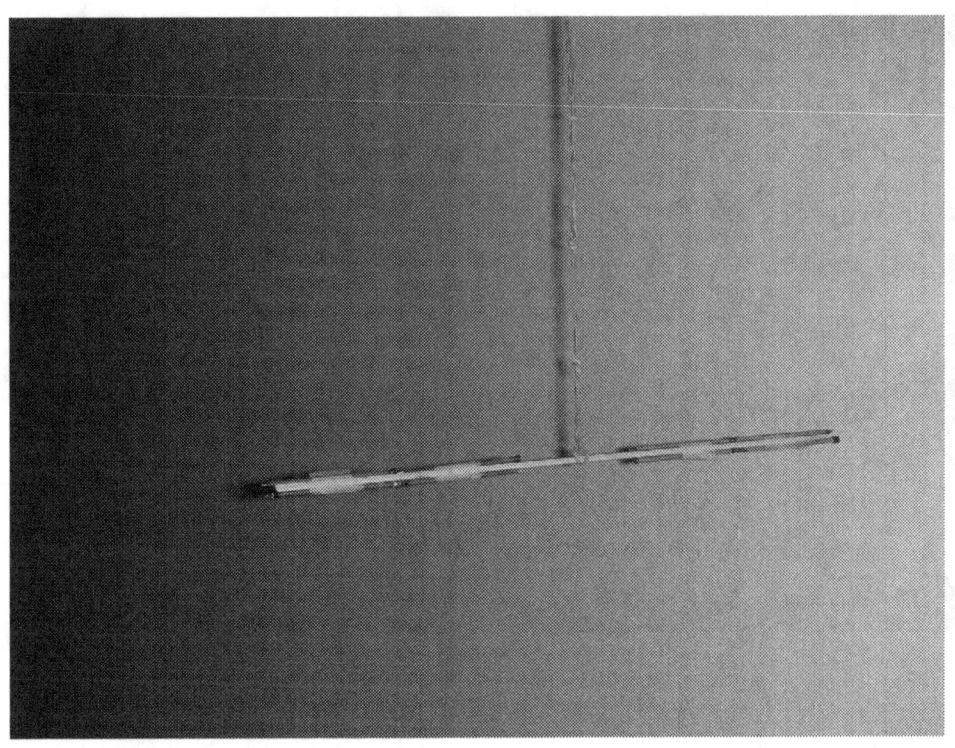

Experiment # 9:Earth's Magnetic field can be determined by measuring the period of oscillation for this magnetic dipole, which moves very slowly after settling down. The period should be approximately 42 seconds (The nails were magnetized using a dry cell battery, but a power supply can also be used if you have one).

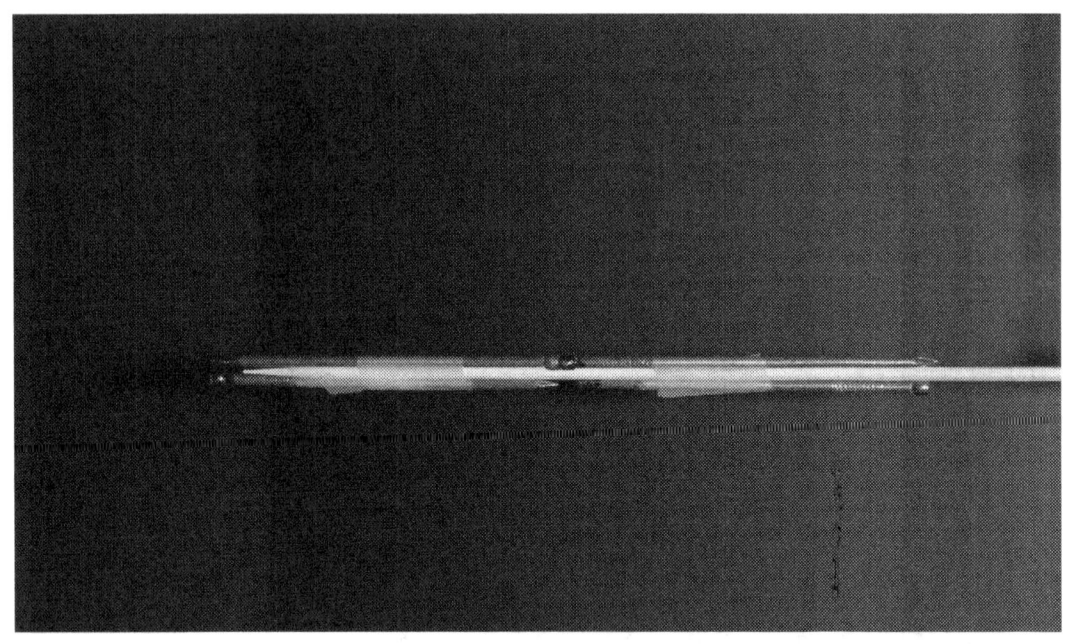

Experiment #9 cont'd)
Close Up of one side of the magnetic dipole used to determine the Earth's Magnetic Field
(The nails were colored black on the North Pole end using a marker. The "North" end
can be determined using the "South" end of a compass needle).

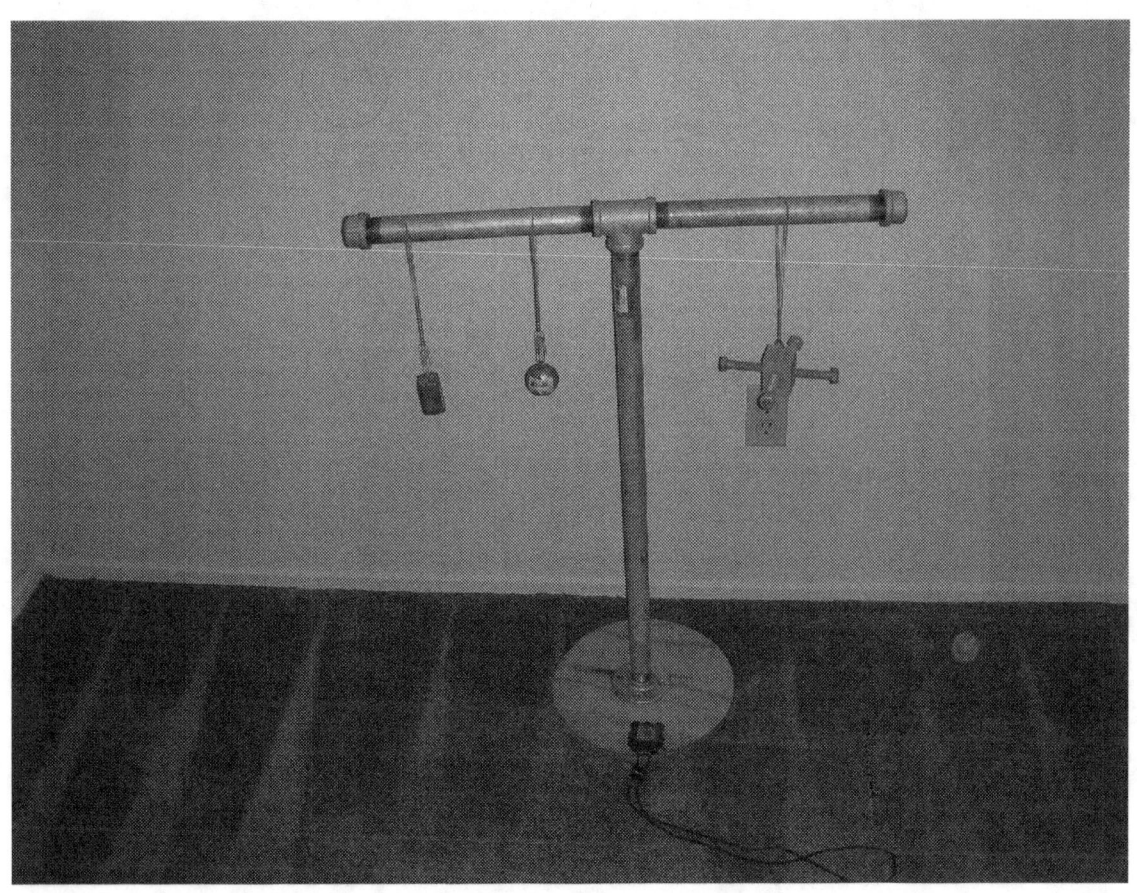

Experiment #3: Harmonic Oscillators and Moments of Inertia

Time their periods of oscillation using any stop watch!

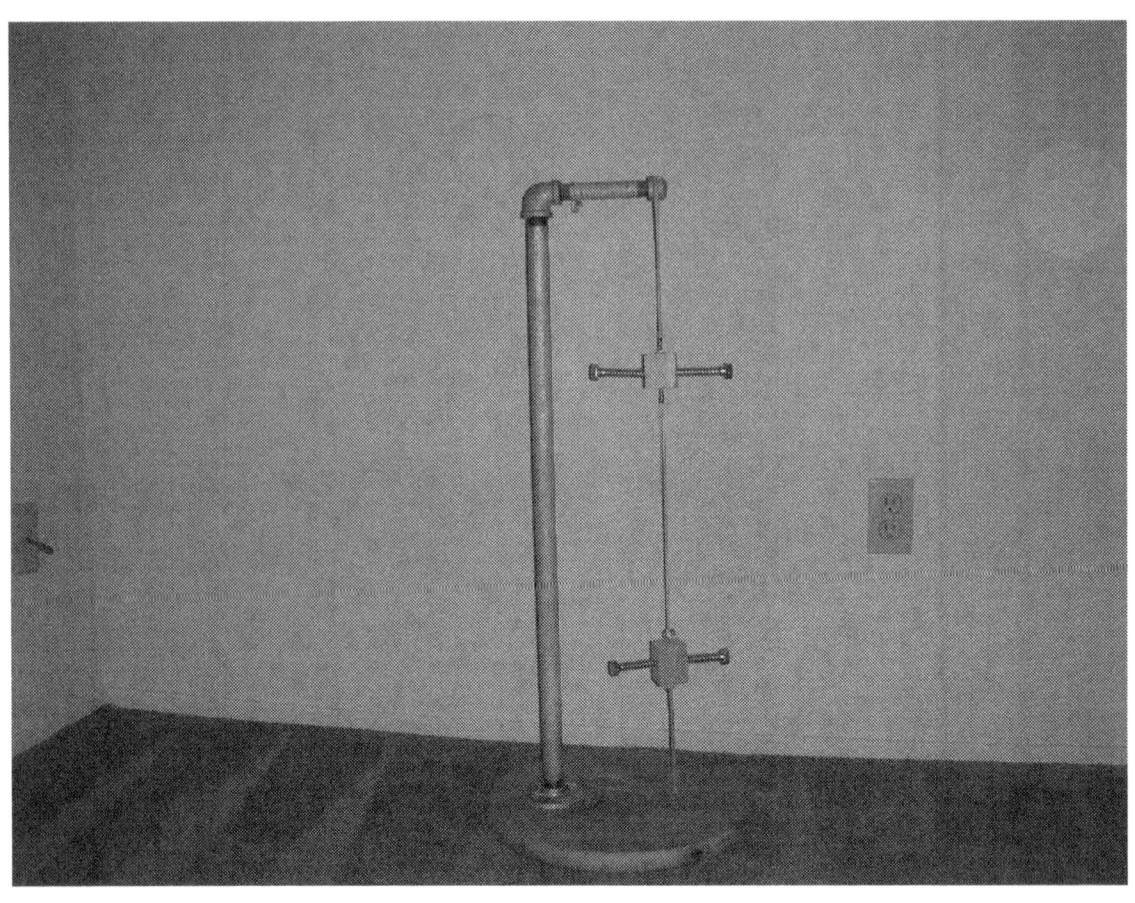

"Coupled Oscillator" found in the Math-Physics Appendix B.

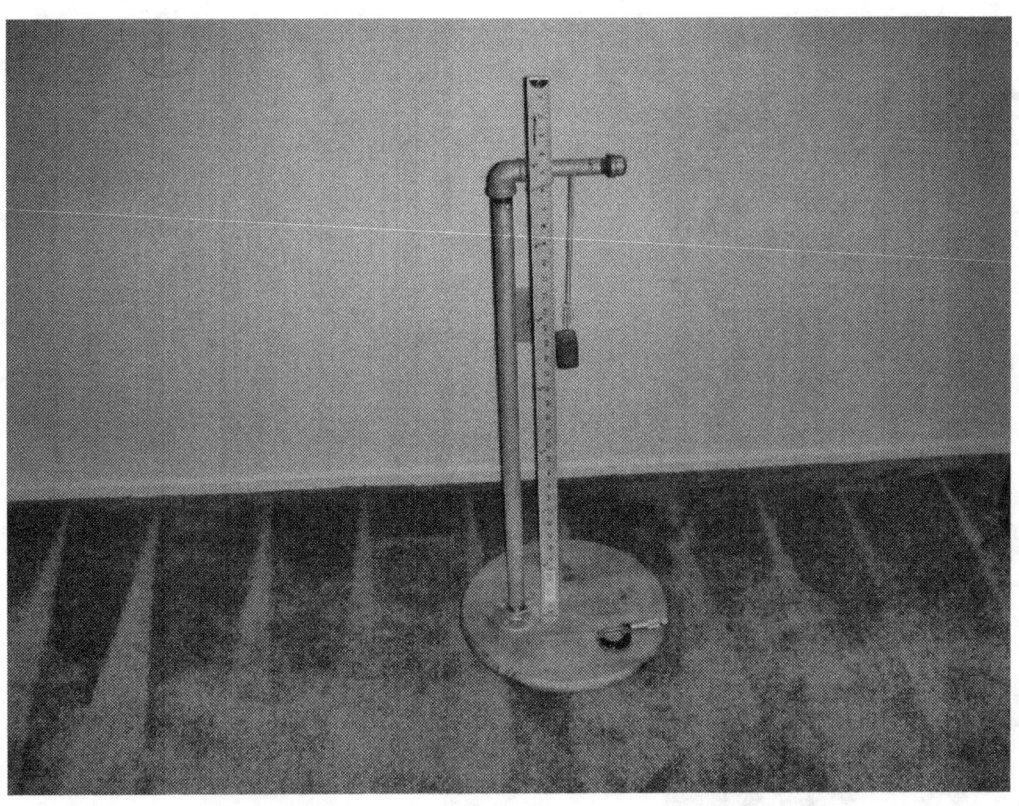

Hookes' Law: Experiment #1
Measure the displacement using the yardstick, by noting the initial point, then each subsequent point for each applied weight.

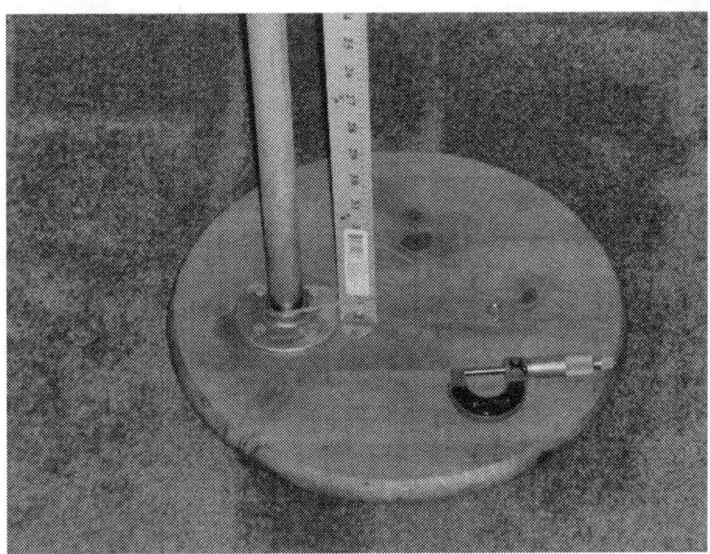

Stress-Strain: Experiment #2
Use a 1" micrometer to measure both the width and thicknesses for each cross sectional area, as more weight is applied (Stress = Applied Force/ Cross sectional Area).

Expansion of the Universe: Experiment #10
See how the galaxies get further apart as the weight applies more "Red shift."

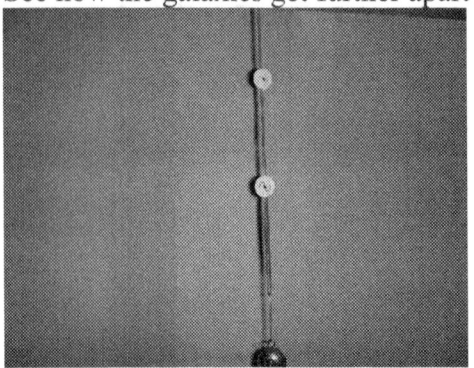

(Red Shift) z = 1.0 (Gyrs ago = 0)

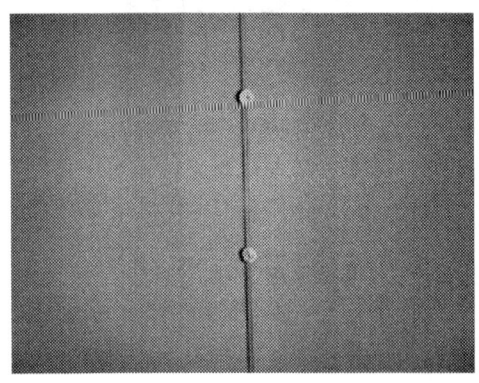

(Red Shift) z = 1.25 (Gyrs ago = 6.5)

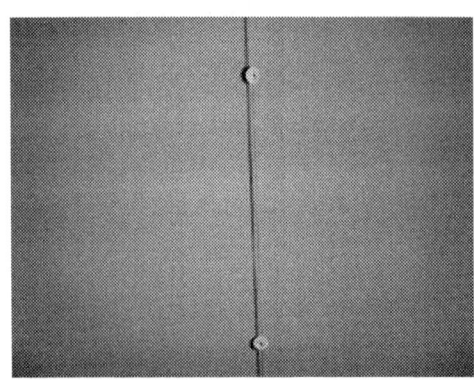

(Red Shift) z = 1.50 (7.7 Gyrs ago)

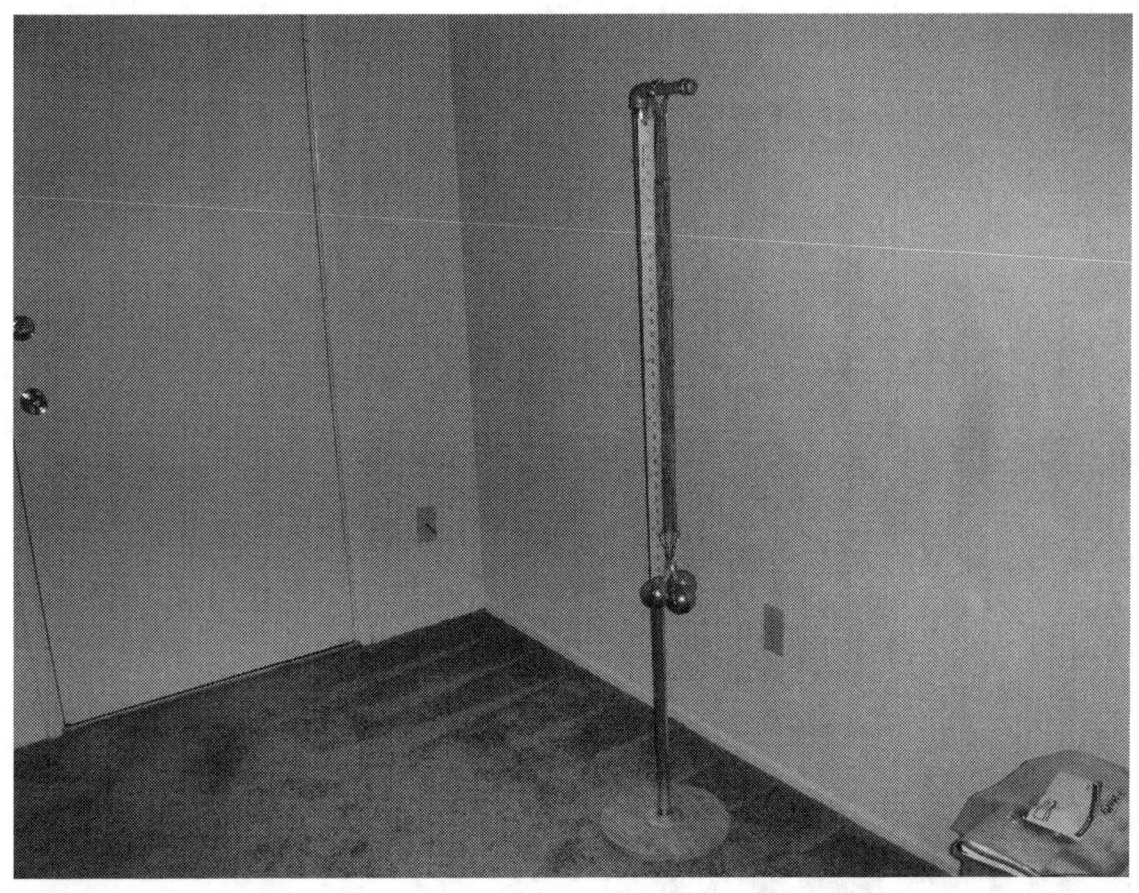

Analog Computer: Experiment #11
11 lbs of applied weight should produce a displacement of approximately 7 inches, to give a value for K/2 (= 11lbs/ 7inches), and a value for K of about π (= 3.1454…).

Close Up of the "measuring point" (choose the "bottom" of the ring)

Microstructures: Experiment #8
Scanning Electron Microscope micrograph of Rubber-modified Epoxy (100nm)
(Thanks to Raymond A. Pearson, reprinted from his paper entitled *"Effect of Rubber-Plastic Zone Interactions on Fatigue Crack Propagation Behavior of Rubber-Modified Epoxy Polymers,"* Journal of Material Science Letters, 1994, 13, p.1460-64.
(Micrographs taken by R. Bagheri, and H. Azimi).

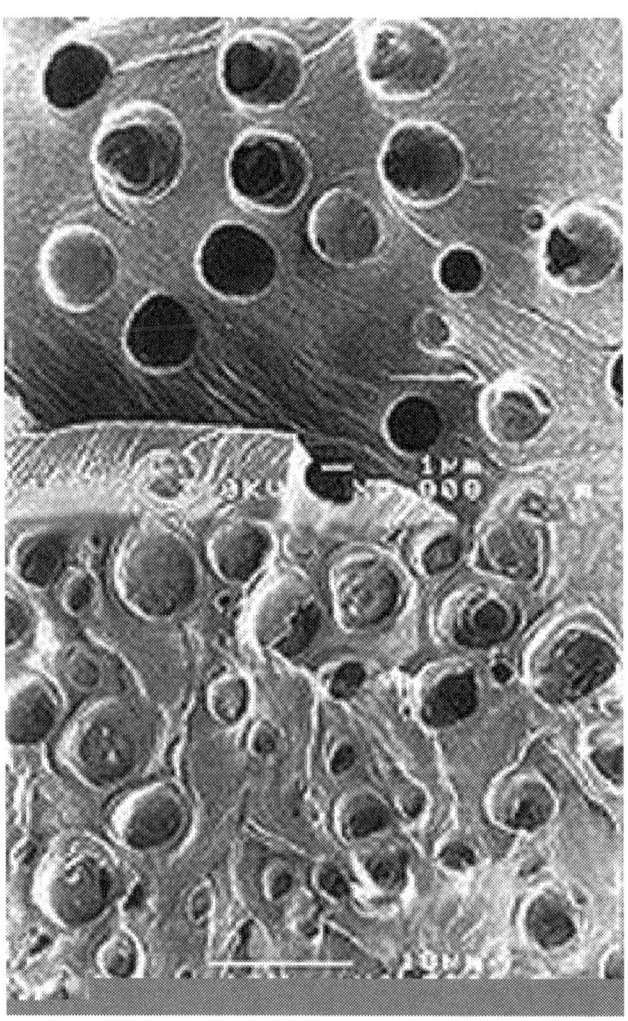

SEM micrographs of fracture surfaces of rubber-modified epoxies: (Top) Static fracture surface (smooth holes) and (Bottom) fatigue crack propagation fracture surface. Note the additional distortion of the matrix seen on fatigue fracture surfaces (Rings of "Wrinkles").

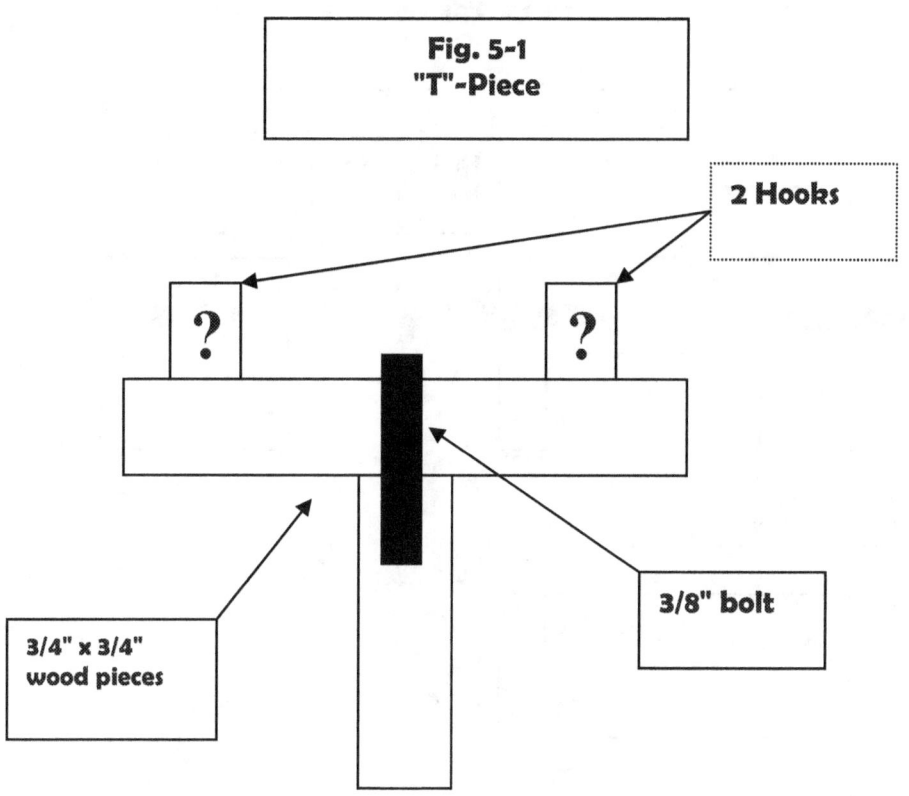

Fig. 5-1
"T"-Piece

2 Hooks

?   ?

3/4" x 3/4"
wood pieces

3/8" bolt

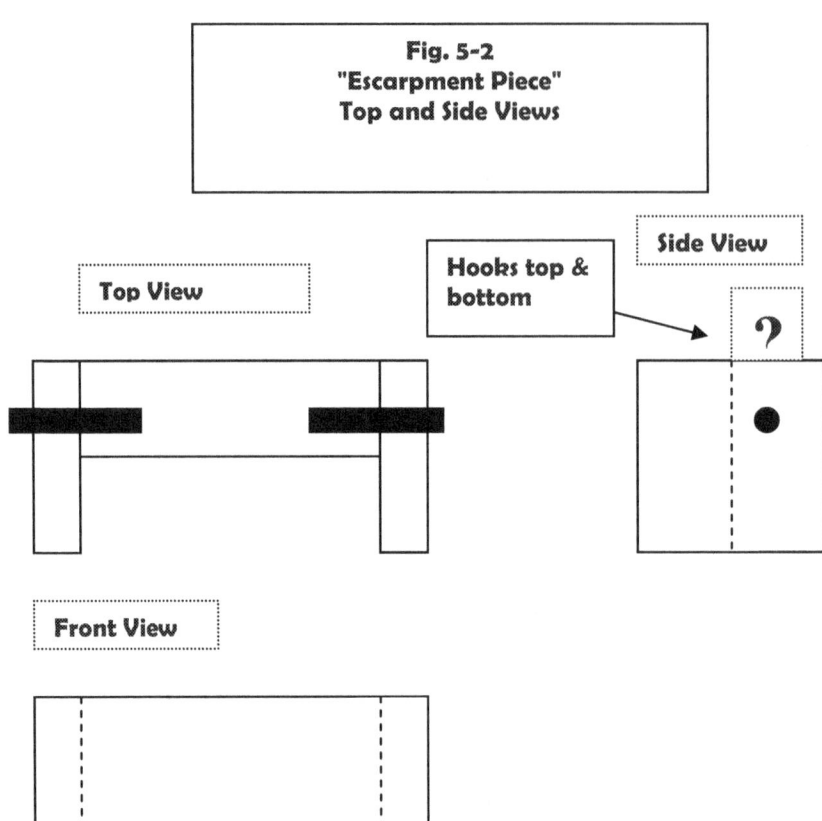

Fig. 5-2
"Escarpment Piece"
Top and Side Views

Top View

Hooks top & bottom

Side View

Front View

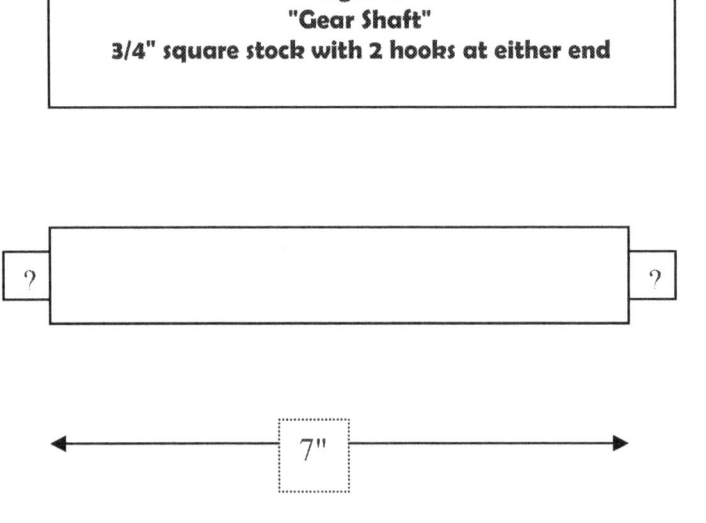

Fig. 5-3
"Gear Shaft"
3/4" square stock with 2 hooks at either end

7"

Fig. 5-3
"Clock Set Up"

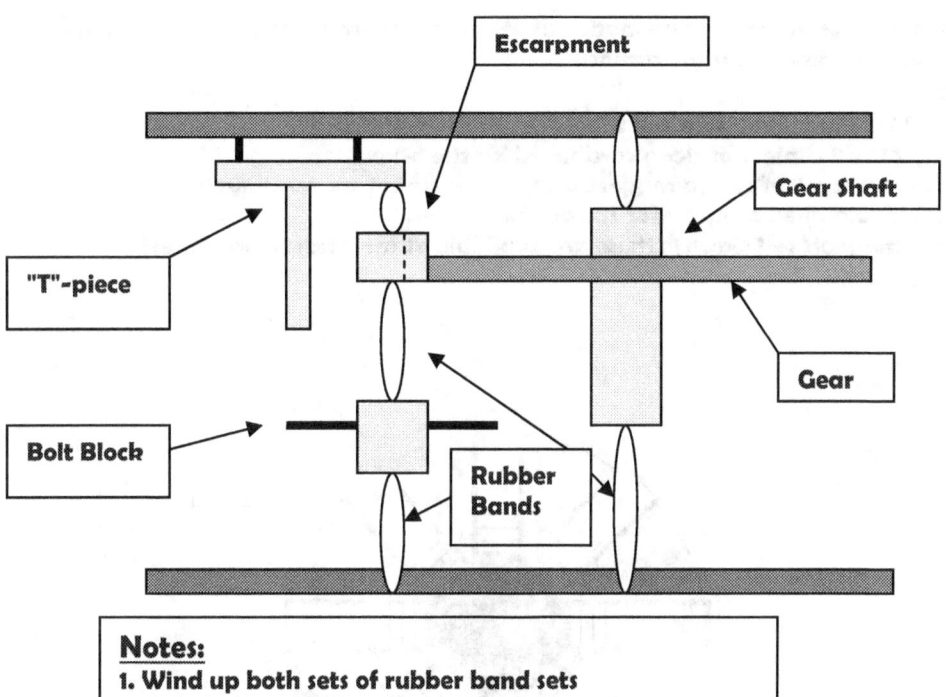

Escarpment

Gear Shaft

"T"-piece

Gear

Bolt Block

Rubber Bands

Notes:
1. Wind up both sets of rubber band sets
2. Adjust gear teeth to fit gear and escarpment

Questions:
1. What is the basic period of the clock?
2. How long does one "wind up" last?
3. Comparing the period of the clock with a stop watch and a pendulum, which measures time more accurately?
4. What are the important applications of such time pieces?
5. What types of energy get converted? When?
6. Albert Einstein wrote a paper in 1905 entitled "Special Relativity" in which he showed that time slows down when the object (with the clock) travels near the speed of light. He explained what happens in his classic "Twins Paradox" where one twin travels near the speed of light, while the other remains at home on earth. When they meet later, the one who stayed home has aged, while the one who traveled near the speed of light in still young!

If you build two clocks and make one of the gears slightly larger or smaller you can see what his effect is like when you run both clocks at the same time! Who is the younger one? Who is older?

Creative Modern Physics Experiments Using Rubber Bands:
Experiment #6: Electric Generator (AC & DC Current, Gyroscope)

A really nice electric generator can be made out of cardboard, magnets, and wire using a rubber band as the source of motion for the armature.

Parts List::
1. (qty 8) Empty toilet paper roll tubes
2. (qty 1) 24" x 24" piece of clean cardboard for the base
3. (qty 1) 100 ft roll of 16awg magnet wire for the armature windings.
4. (qty 1) Liquid filled compass for the galvanometer.
5. (qty 4) magnets (~ 1 Gauss) (these are available at most hardware stores)

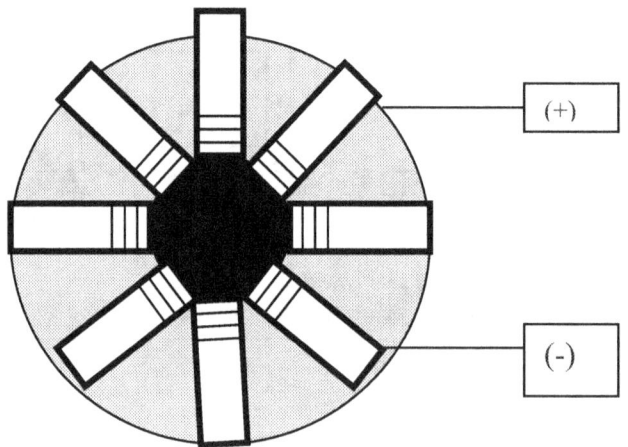

Fig 6-1
Electric Generator:
Tubes with Coils(100 turns) tapped onto Base Assembly

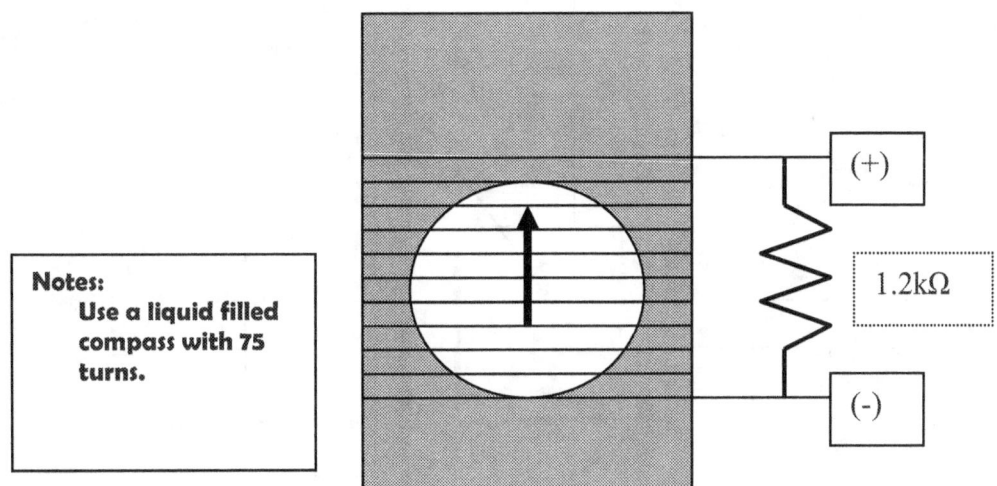

Notes:
   Use a liquid filled
   compass with 75
   turns.

(+)

1.2kΩ

(-)

Fig 6-2
Galvanometer for measuring the current from the generator

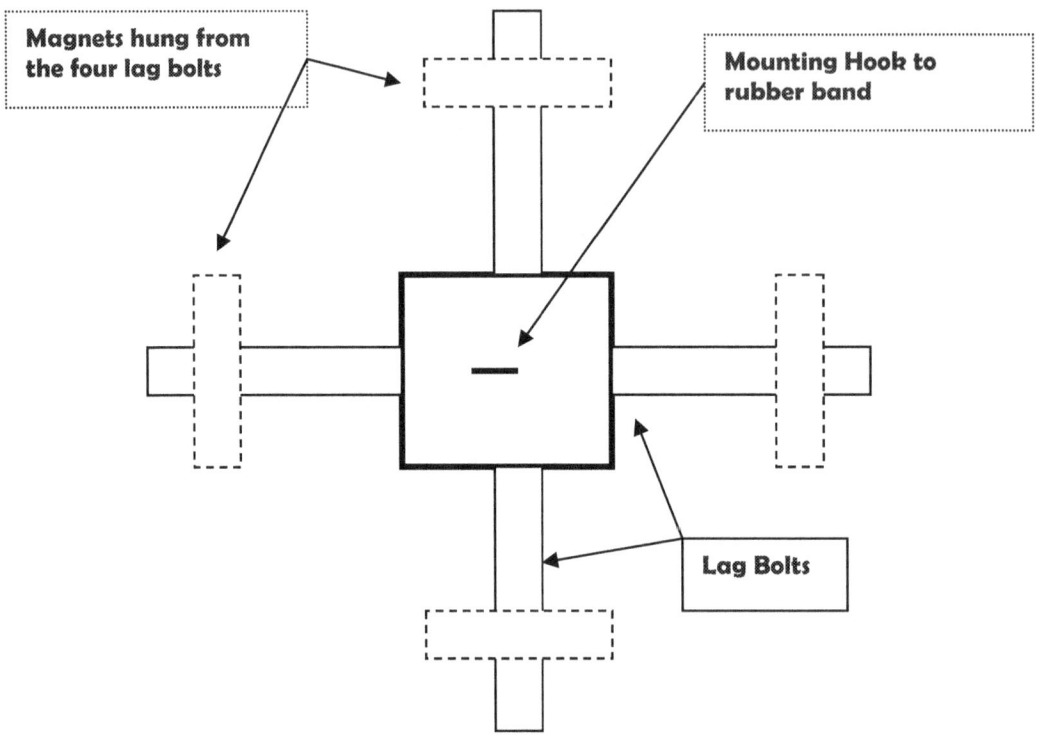

Magnets hung from the four lag bolts

Mounting Hook to rubber band

Lag Bolts

Fig 6-3
Armature: Top View
Showing Block, four Lag Bolts, the mounting hook, and the locations for four magnets. With this setup it is possible to measure the fluctuations in the compass needle, as well as to observe the "gyroscope effect."

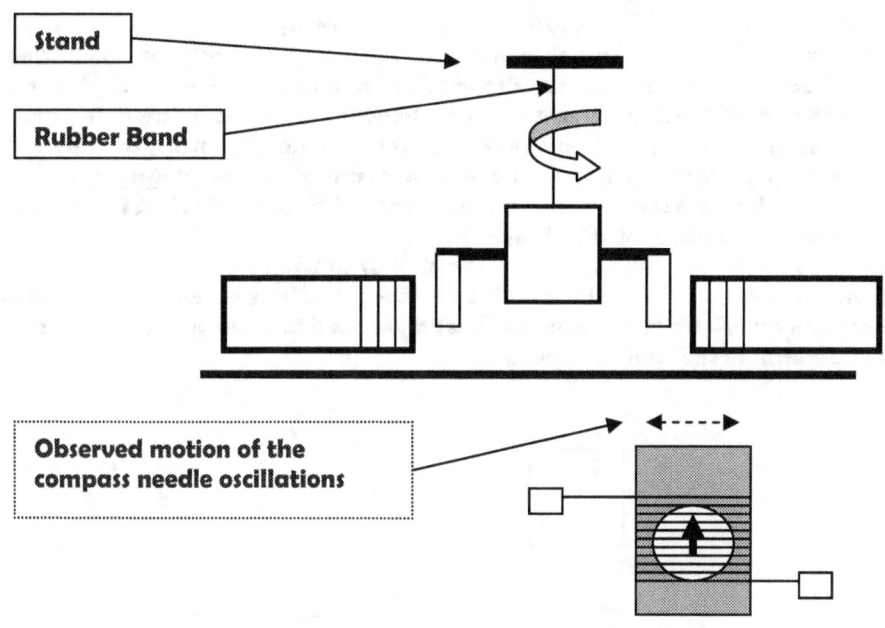

Stand

Rubber Band

Observed motion of the compass needle oscillations

```
Fig 6-4
Final Set Up
Showing Armature Assy rotated by the wound rubber band
so the magnets induce current into the coils.
```

Theory:

1. **The Electromotive Force (EMF) is given by**
    $\mathcal{E} = -d\varphi/dt = d(BA \, Cos\theta)/dt \quad dA/dt = Blv$

2. **the Magnetic Flux**
    $\Phi = BA \, Cos\theta$, where θ is the angle between B and the normal vector to the plane

3. **Lorentz Force**
    $F_b = qv \times B$

4. **Torque**
    $T = \mu \times B$, where $\mu = NIA$, N the number of turns, I the current, and A the area.

Questions:

1. How do you know when you're generating electricity? What kind is it?
2. How does the generator also behave like a gyroscope?
3. What is the value of the Electromotive Force in the system?
4. How much Torque is there in the B-field?
5. How can the magnetic induction; B be measured?

Experiment #7: Gravity Waves

This sounds like a complicated thing, but it turns out to be quite easy to do!

While reading the "New York Times," I came upon an article about radio astronomy entitled *"Refining the Art of Measurement"* by <u>Malcolm Browne</u>, wherein he describes the work done by a group of radio astronomers at Arecibo, Puerto Rico who believe they have isolated and measured "Gravity waves" as the phonons given off by a pair of co-rotating neutron stars, their period of rotation was measured by <u>Dr. Joseph Taylor</u> and one of his graduate students <u>Russell A. Hulse</u>, and it was determined to be 0.06 seconds. This period turns out to be equivalent to a frequency of 104.7 Hertz (the acoustical branch of the phonon!).

The physical system of two co-rotating neutron stars is similar to what physicists call a "rigid rotator" – the eigenvalues for the simple harmonic oscillator (like those given in the previous section on phonons) don't work for this model! So instead we need to develop a different set of relations to help us determine the "phonon spectra."

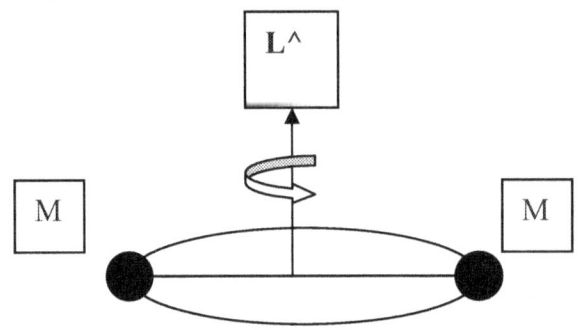

Fig 7-1
Model for the "Rigid Rotator" analog for the co-rotating neutron stars with radius "a"

Theory:

1. The Moment of Inertia for the system is given by $I = 2Ma^2$
2. The Energy for this system is $E = L^{\wedge 2} / 2I$, where $L^\wedge$ is the angular momentum operator.
3. Then the time-Independent Scrödinger is $H^\wedge \varphi = (L^{\wedge 2}/2I)\, \varphi = E\varphi$
4. We can rewrite the this equation in terms of the eigenvalues of $L^{\wedge 2}$ namely l & m.
   $(L^{\wedge 2}/2I)\, \varphi_{L\,m} = E_L\, \varphi_{L\,m,}$, $L^{\wedge 2} = h^2 L\,(L+1)$
5. And finally, we get the correct relation to describe the eigenvalues of the phonon
   $E_l = h^2 L\,(L+1) / 2I$
6. We can now graph the expected harmonics from the phonon spectra.
7. For the case of angular harmonic motion and the "rigid rotator," the eigenvalues (the component frequencies of the system) have a "parabolic envelope" and this is what the radio astronomers measured.

8. $L = I\omega = 2Ma^2\,(2\pi/T)$

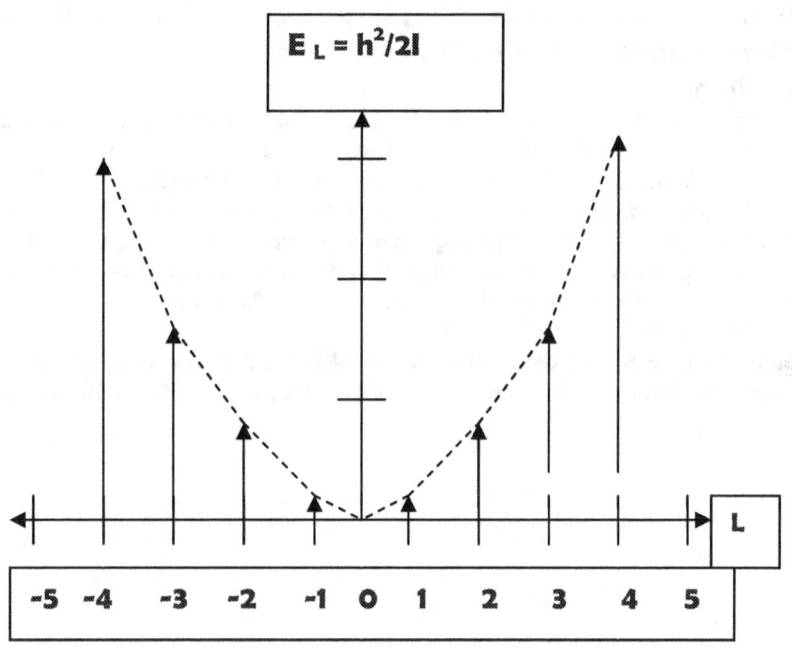

Fig 7-2
Phonon Spectra for the Co-rotating Neutron Stars

Using the same angular harmonic oscillator as was used in the Harmonic Oscillator experiment, imagine that the bolts are the co-rotating neutron stars and time their rotational periods just like the radio astronomers at Arecibo did. Use this fundamental period to estimate the exact values for the phonon spectra.

Experiment #8: Microstructures (The "*chaotic*" inner world of rubber)
Microstructures and Micromechanics:
Plastics & Polymers

Polymers large organic molecules made by nature, usually in the hydrocarbon series, $C_m H_{2m+2}$ with melting temperatures related to their molecular size.

Microstructures are those characteristics of a material which define the morphology between phases in equilibrium (gas, liquid, or solid) of differing composition and are usually shown in scanning electron microscope (SEM) micrographs as phase boundaries, grain sizes, grain growth, crystallization, or polymorphic reactions in pure materials. Some of the more important physical characteristics of a material might include: optical index, Thermal conductivity, Thermal expansion, and strength.

<u>Polyisopropene</u> is a natural rubber coming from rubber trees and plants mostly found in the rain forests of South America and Africa. It has a "Vinyl" structure $H_2CCH{:}CRCH_2$ as is shown below

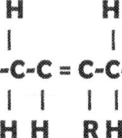

Fig 1. Vinyl structure

Where the R is a general term denoting a "side group" such a propylene (CH_3), then the structure is $H_2CCH{:}CCH_3CH_2$

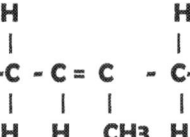

Fig 2. polyisoprpene

The following data are for Polyisopropene:

Density (ρ)	0.91 Mg/m³
Young's Modulus	0.002-0.1 GPa
Tensile Strength	~ 10 MPa
Glass Temperature (T_g)	220 °K
Softening Temperature (T_s)	350 °K
Specific Heat (C)	2,500 Joules/Kg °K
Thermal Conductivity	0.15 W/m °K
Thermal Expansion Coefficient	600 M/°K

<u>Molecular weight of Polyisopropene monomer:</u>
 C = 12g/mol x 5 = 60
H = 1 g/mol x 8 =   <u>8</u>
                68 g/mol per monomer

Micromechanics:
Is the expression of the mechanical properties for very large macromolecules.

1.   Mass average molecular size : $M_m = \sum (W_i M_i)$
      Where W_i = mass fraction in each size interval
             M_i = representation value (middle) of each interval

2.   Degree of Polymerization (DP):

      DP= molecular mass (amu/molecule) / mer mass (amu/mer)

3.   Molecular weight = DP x molecular weight of monomer = DP x m

4.   Avg molecular weight = $\underline{DP}$ = $\int_0^\infty DP\ p(DP)\ d(DP)$, where p(DP) is a normal distribution
      In other words the average degree of polymerization is the expectation value of the normal distribution of the degree of polymerization p(DP)
      $P(x = DP) = p(x; \mu, \sigma) = (2\pi)^{-\frac{1}{2}} \sigma \exp[-(x - \mu)^2 / 2\sigma^2$
      The standardized variable is $Z = (x - \mu)/\sigma$
      Then
      $P(a < DP < b) = P[(a - \mu)/\sigma < Z < (b - \mu)/\sigma] = \Phi[(b - \mu)/\sigma] - \Phi[[(a - \mu)/\sigma]$

5.   Mass average molecular size : $M_m = \sum (W_i M_i)$
      Where W_i = mass fraction in each size interval
             M_i = representation value (middle) of each interval
6.   "Number Average" molecular size M_n is given by
      $M_n = \sum (X_i M_i)$
      Where X_i = the numerical fraction of molecules in each size interval
             M_i = the representative (middle) value in each interval

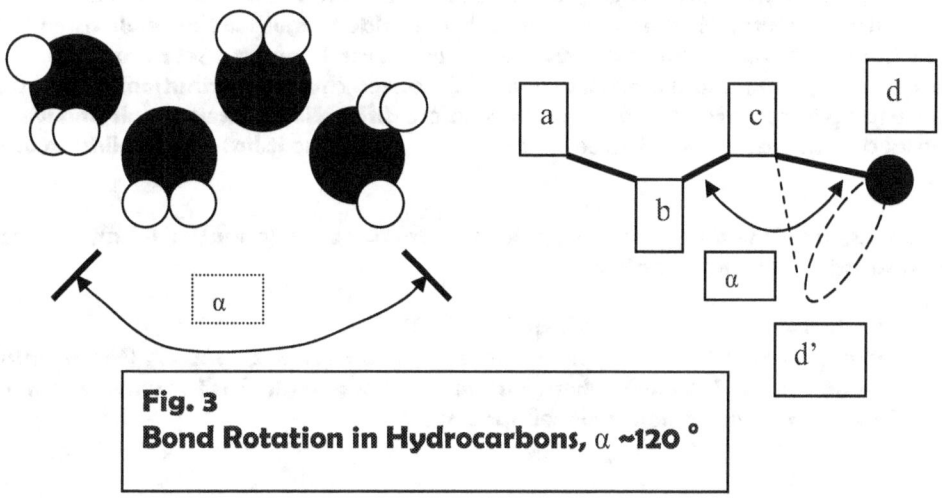

Fig. 3
Bond Rotation in Hydrocarbons, α ~120 °

7. Molecular Length of a polymer (the root mean square length) ("Random Walk") = $\lambda \sqrt{n}$
= [the chain spacing (λ)] times [The square root of the number of "steps" (n)]

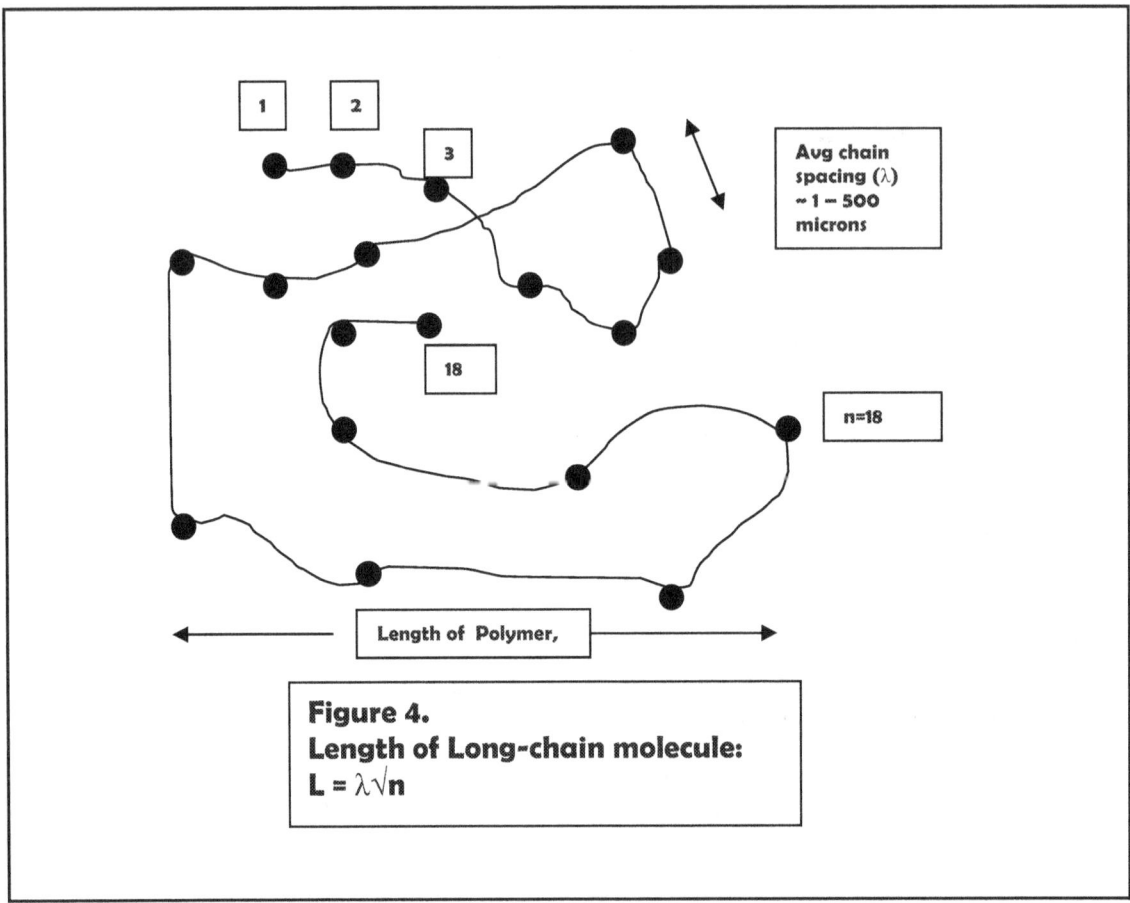

Avg chain spacing (λ) ~ 1 – 500 microns

n=18

Length of Polymer,

Figure 4.
Length of Long-chain molecule:
$$L = \lambda \sqrt{n}$$

Amorphous Unit Cells:

Even though a polymer such as polyisopropene is considered to be "amorphous," that is lacking any definite structure, when enough mer units are added together the whole assemblage works some what as if it were a molecular crystal, although really being a set of random, interconnecting, long-chain molecules, with delocalized charge distributions is truly what it is. When x-ray powders are performed an incomplete diffraction pattern result and we say the material demonstrates some degree of "crystalinity" because it has some ability to diffract x-rays.

The response from the diffracting x-rays follows Bragg's Law (eauation 8.) and approximates a Sinc (x) function as is shown below.

8. $2d \sin \theta = n\lambda$ (Bragg's Diffraction Law),
where n = ±1, ±2, ±3, . . . (for the maxima) and ±1/2, ±3/2, ±5/2, . . . (for the minima)

9. $\theta = \sin^{-1} \{n\lambda/2d\}$ (theta), where θ is the scattering angle, λ is the wavelength of the x-rays, and d is the amorphous unit cell spacing.

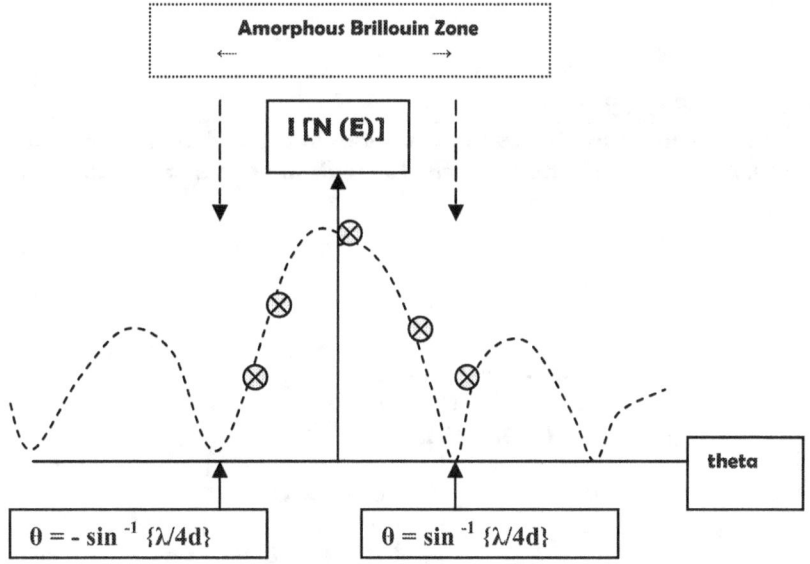

Figure 5A & B.
Data from an x-ray powder, visible light, or TIR test done on a polymer having an amorphous structure, showing that the response improves with an increased number of data points. The amorphous Brillouin Zone is shown
(within n = ± ½)

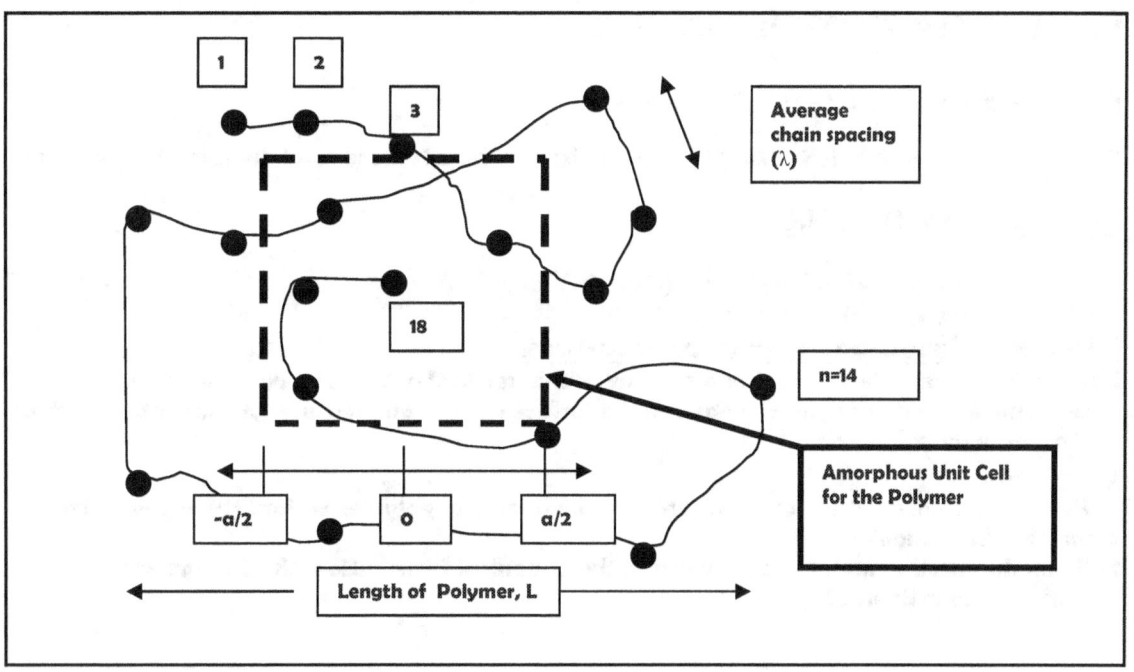

Cross Linking and Branching:

This occurs whenever bonds form between the various interlaced long-chain molecules and help to build strength, but if their numbers become too high they often also build "brittleness" into the material.

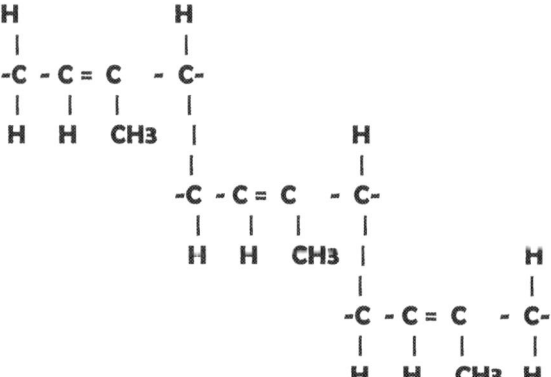

Fig. 5
Cross Linked polyisopropene molecules

Canonical Distribution Function for the Simple Harmonic Oscillator:

10. $\rho_c = C \exp \{ -V/kT \} = C \exp \{ Kx^2/2kT \}$, where C is the Specific Heat

Configurational Entropy for the S.H.O. :

11. $.\Delta S = Nk [1+ \ln \{ k \Delta T/ N \hbar\omega \}$ (of time)

12. $S (\omega, T) = < -k \ln (-Kx^2/2kT)^N \exp (-Kx^2 2kT) >$, where N is the total number of oscillators.

Free Energy for the S.H.O. :

13. $F (x,T) = -kT \ln Z = -kT \ln \{-(Kx^2/ 2kT)^N \}$
14. $F(\omega, T) = -kT \ln \{ (\hbar\omega/ kT)^N \}$

Microscopic Investigation of Rubber Surface:

It is possible to take the rubber band and cut a "thin section" out of it to examine under a microscope. If you do this you can observe the surface texture and features of the rubber itself, as well as draw them.

Questions:

1. Given that $\mu = 0$, $\sigma = 0.10$, can you compute the degree of polymerization? (hint: you will need a normal distribution).
2. Given the Specific Heat, C; can you graph the Canonical Distribution, the Configurational Entropy, the Free Energy?

Properties of Selected Materials:

Material	Density (g/cm^3)	Thermal Conductivity (W/°C/mm)	Linear Expansion °C^{-1}	Electrical Resistivity Ohm*m	Average Modulus (MPa)	Average Modulus (psi)
Metals:						
Copper (99.9+)	8.9	0.40	17×10^{-6}	17×10^{-9}	110,000	16×10^6
Monel (70 Ni/30 Cu)	8.8	0.025	15×10^{-6}	482×10^{-9}	180,000	26×10^6
Silver (sterling)	10.4	0.41	18×10^{-6}	18×10^{-9}	75,000	11×10^6
Ceramics:						
Al$_2$O$_3$ (Alumina)	3.8	0.029	9×10^{-6}	$>10^{12}$	350,000	50×10^6
Glass:						
Plate	2.5	0.00075	9×10^{-6}	10^{12}	70,000	10×10^6
Borosilicate	2.4	0.0010	2.7×10^{-6}	$>10^{15}$	70,000	10×10^6
Polymers:						
Rubber (synthetic)	1.5	0.00012	------	-------	4-75	600-11,000
Rubber (Vulcanized)	1.2	0.00012	81×10^{-6}	10^{12}	3,500	0.5×10^6
L.D. Polyethylene	0.92	0.00034	180×10^{-6}	$10^{13} - 10^{16}$	100-350	14,000-50,000
H.D. Polyethylene	0.96	0.00052	120×10^{-6}	$10^{12} - 10^{16}$	350-1,250	50,000-180,000
Polystyrene	1.05	0.00008	63×10^{-6}	10^{16}	2,800	0.4×10^6
Polytetrafluoroethylene (Teflon)	2.2	0.00020	100×10^{-6}	10^{14}	350-700	50,000-100,000
Nylon	1.15	0.00025	100×10^{-6}	10^{12}	2,800	0.4×10^6

* This data taken from Van Vlack (1980) "Elements of Materials Science and Engineering," p.522-3

Experiment #9: Earths Magnetic Field (fields and plasmas)

Parts List:
1. (qty 1) Chop stick ~ 20 cm long.
2. (qty 8) 2" finishing nails ("magnetized"- See Appendix D).
3. (qty 1) paper clip
4. (qty 1) rubber band
5. (qty ~ 24 inches of) sewing thread.

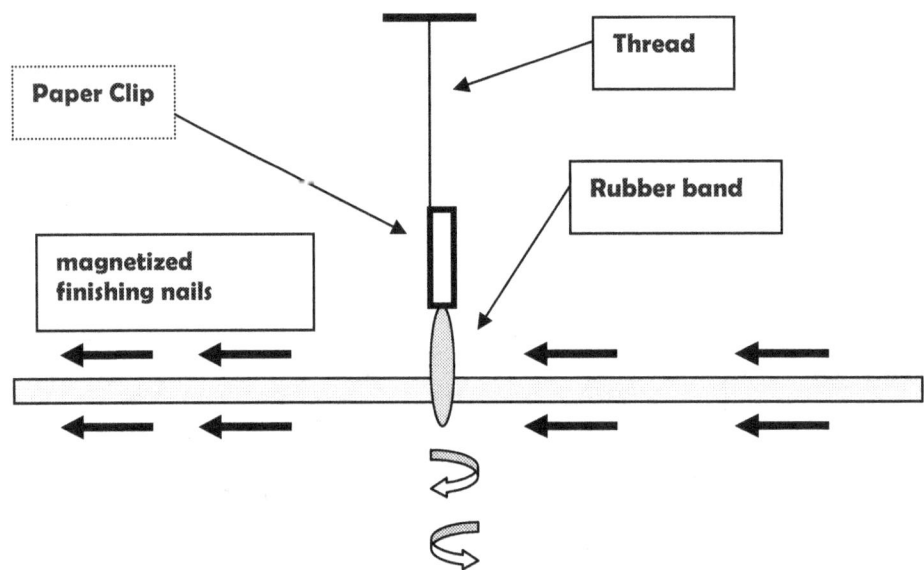

Fig 9-1
"Chopstick" with eight "magnetized" finishing nails.
The period of oscillations can be measured using a
stopwatch when the chopstick is "gently" set into motion.

Equations:

1. Lorentz Force: $F = qv \times B$, or $F = qvB \sin\theta$
2. Sum of the Torques is zero!
3. $\tau\,net = I\,\theta'' = \Sigma\,|\,r \times F| = \tau + \tau = I\,\theta'' + rqvB\sin\theta = 0$
4. Small θ approximation: $\sin\theta \sim \theta$
5. For the Simple Harmonic Oscillator:
   SHO: $\theta'' + \omega_0^2\theta = \theta'' + (rqvB/I)\,\theta = 0$
6. Comparing with the standard form of the simple harmonic oscillator:
   $$\theta'' + \omega_0^2\,\theta = 0$$
   Leads us to conclude that $\omega_0^2 = rqvB/I$, where m is the mass of the nails on one side (~2.5 gm), q = 1esu, and $v = r\omega_0$.
7. Therefore the Earth's magnetic field may be determined by solving for B:
   $B = \omega_0^2\,I/rqv = \omega_0\,m\,r^2/2q = \omega_0\,m\,/2\,q = \pi m/qT$ and so the final form becomes

   $B = \pi\,m\,/\,q\,T$, where q = 1 esu because of the oscillating magnetic dipole interaction, $m \sim 6.7 \times 10^{-5}$ Kg, T is the measured period of oscillation, and B will be measured in Teslas (T).
   $B = \{\pi(\,6.7 \times 10^{-5}\text{ kg})\,/\,(1\text{esu})(\,42.207\text{ sec}) = 0.4987 \times 10^{-5}$ T

This turns out to be a very sensitive and accurate way to measure the Earth's magnetic field. Try it for a value of T equal to approximately 42.207 sec as measured by a stopwatch and timing 10 periods.

Though the actual value will vary with location and also over time, the accepted value given in most books is for the magnetic induction (B) is approximately
0.5294×10^{-5} Teslas.

Questions:

1. What is the source of earth's' magnetic field? How does the "Bow Shock" between the earth's magnetic field and the "solar wind" help to reduce the amount of radiation received by the earth's surface? What would happen if the earth had no magnetic field?
2. Since you have already measured the earth's magnetic induction, can you also determine its magnetic strength?
3. How is this experiment a good example of "magnetohydrodynamics?"

Experiment #10: Expansion of the Universe (Hubble's Law & galaxy clustering)
(The following is an excerpt from a paper prepared by the author, David Tracy; for a course at Cal-State Long Beach)

"Biography of Edwin Hubble"

Introduction:

For my biography of a famous scientist I choose the American Astronomer Edwin Hubble, who is credited with being the first scientist to discover that the universe is expanding. He based this conclusion on a survey of the red shifted spectra collected from many galaxies. He proposed a theory called "Hubble's Law" which relates the velocity of a receding galaxy with its distance. The mathematical constant of expansion, which bears his name is "Hubble's Constant." It also gives a rough estimate of the age of the universe. Many years ago as an undergraduate, I computed this "age" to be about 16-18 Gy ago. Edwin Hubble is an impressive individual because of the scope of his discoveries and their impact even today on the fields of astrophysics and cosmology. Though he had humble beginnings, he excelled in school and won many important scholarships and awards, which helped propel his career. His story begins in the Midwest town of Springfield, Missouri.

Biographical Information:

Martin Jones, the grandfather of the Edwin Hubble the astronomer, moved to Springfield, Mo in March of 1906. Their family traced their roots back to Justice Hubbell, a signer of the "Articles of Association" protesting the closing of the port of Boston in 1775. He had family members who served on both sides of the Revolutionary war, the war of 1812, U.S./Mexico war, and the Civil war. John Powell Hubble and Mary Jane Hubble would marry and have seven children, the oldest of which was Edwin. Family tradition dictated that the first son of a Hubble receive special consideration in terms of education and moral guidance. Raised as a Baptist, his father had grown up during the era of "Reconstruction" in the south, Giles County, Tennessee, where his own father (Edwin's grand father) had been a slave holding plantation owner. The Missouri of Edwin's childhood was a semi-southern atmosphere, characterized by its "river life." Among Edwin's favorite books as a child was "Tom Sawyer" by Mark Twain another fellow Missourian. He was also fond of the books by another writer; Jules Verne, which described how man might travel to the center of the earth, or to the bottom of the sea.

Edwin grew up during the era of "Prohibition." He once entertained his young friends by informing them about a total eclipse of the moon after midnight on June 23, 1899. He stayed outside to watch and not miss any of it!

He entered school in September of 1901, and he excelled in many subjects. He graduated "with honors" and won a scholarship to the University of Chicago, which at the time was the finest school west of New England. The faculty included Albert Michelson as head of the physics department (Michelson was to later play a role in the discovery of "Relativity." The Michelson-Morley Experiment established that the speed of light is constant in all directions, a fact that helped a young physicist Albert Einstein to develop his theory of "Special Relativity"). While he was in college he liked to listen to "Rag" music, joined a campus musical group called the "Black friars," and learned to perform "magic" tricks. His undergraduate days were a time of great financial stress. He only received about half of the scholarship money he was promised. He did brilliantly in his studies though and regretted not being able to take more courses in mathematics.

In that freshmen year, he had his first serious romance with a girl named "Elizabeth," it would be his only one until he married many years later.

His sophomore year ended June 9, 1908 and he received a Jr. College Scholarship in Physics, this allowed him to become Robert Millikan's research assistant (Millikan's famous "oil drop" experiment provided the first estimate for the charge-to-mass ratio for the electron).

He won a "Rhodes Scholarship" in 1910 and got to study law at Oxford where he became captain of their baseball team.

He went to Yerkes Observatory, in Chicago, in 1914 as a research assistant.

During the span of years from 1917-19 he served as a captain in the infantry during WW1.

After the war he returned home to Springfield where he practiced law, taught high school physics & mathematics, and also couched the basketball team.

During a visit to the University of Chicago, Walter Adams met a graduate student named <u>Edwin Hubble</u>, who was putting the finishing touches on a dissertation entitled "Photographic Investigations of Faint Nebulae." Adams, representing George Ellery Hale, who was building a new 60-inch telescope at Mount Wilson in California, offered the newly minted Dr. Hubble a job at a salary of $1,200/mo, to become one of the observatory's first five researchers. This event opened the door for all of his subsequent discoveries. In the year 1923 he used the Cepheid variable stars in the Andromeda galaxy (M31) to prove that this "spiral nebulae," as it had previously been thought to be, was in fact, beyond the borders of our own Milky Way galaxy. In that same year he proposed his famous conclusion about the expansion of the universe, and an estimate for the age of the universe as a consequence. He, along with his assistant Milton Humason, devised a scheme for the classification of galaxies that is still in use today.

Between 1923 and 1929 he added 23 more to a growing catalog of galaxies, their red shift spectra, and estimates for their distances. He added estimates for another 21 galaxies, this helped put his velocity-distance relation; "Hubble's Law, on a more secure experimental footing.

Scientific Achievements:
1. *Compiled photographs of many galaxies and created the first "Classification" scheme to distinguish between the various types of galaxies.*
2. First astronomer to use the "Cepheid Variable" technique to estimate the distance to an object outside the Milky Way.
3. First to measure "red shifts" of the spectra of galaxies and established the experimental evidence for "Hubble's Law."
4. Created the first estimate of "Hubble's constant (which turns out not to be a constant! New work in this area of cosmology has proven that the expansion of the universe is accelerating!)
5. First to estimate the age of the universe.
6. He was involved in the construction and completion of the 200-inch Hale telescope on Mount Palomar, which for most of the 20th century was considered the finest research telescope in the world! Now we have the Hubble Space Telescope (HST), which bears his name, and can perform a variety of astronomical measurements high above earth's atmosphere 24 hours a day.

<u>Hubble's Law:</u>
Is a the mathematical description for the relationship between a galaxies "red-shift" and its distance away from us. Luckily for us, this happens to be a linear relationship like the ones we have already explored in this book ("Hooke's Law, etc). To perform this experiment all we need is a couple of friends to hold the ends of a chain of rubber bands looped together, and something to measure the distances between as we stretch them. We can even use some stick-on dots to represent the galaxies. As we stretch the rubber band we can see how the spacings between the galaxies changes. If we do it correctly, our spacings should agree with Hubble's Law.

Let the strain represent the redshift of a galaxy:
(10-1) $z = (\lambda - \lambda_0)/\lambda_0 = \Delta\lambda/\lambda_0$
Let the stress represent the" speed of recession"
(10-2) $v = zc = Hr$, c is the speed of light and H is Hubble's constant = 50 km/s/Mpc
Now we can repeat the "stress-strain" experiment like before except that this time we will think of it in terms of weights being recessional speeds and the strains as redshifts, while spring constant of the rubber band is Hubble's Constant.

Have two people each hold either end of a long rope of looped rubber bands. Put two gum label "Galaxies" on the rope. Have each end holder take a step back, then another, and another making strain measurements as you go. Graph the data to obtain the curve for "Hubble's Law."

Experiment #11 Approximating π using an Analog Computer

Back in the section about Robert Osserman's book, "the Poetry of the Universe," we mentioned that the writer Sarcobosco (John of holywood) in his book "*The Sphere*" made use of the number "pi" to determine the Earth's diameter which he approximated using the ratio 22/7. In this experiment, we make a computer out of rubber bands to help us show this.

- ## Serial Connections of Rubber Bands:

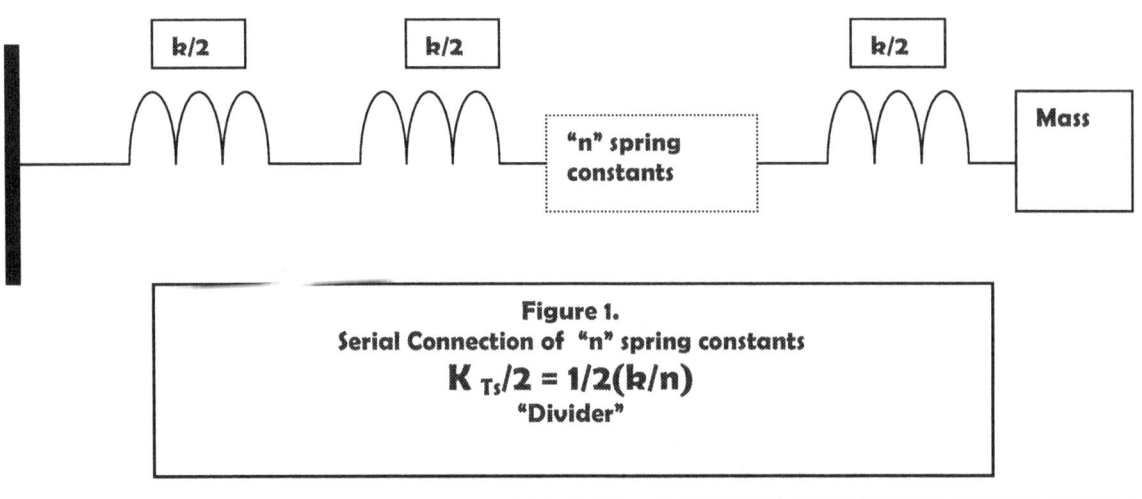

Figure 1.
Serial Connection of "n" spring constants
$$K_{Ts}/2 = 1/2(k/n)$$
"Divider"

The serial connection of "n" spring constants has a combined spring constant, K_{Ts} which can be computed using the following

1. $2(1/ K_{Ts}) = 2\{(1/k) + (1/k) + * * * (n) * * * +(1/k)\}$

2. Or $K_{Ts} = k/n$ (every time you loop two rubber bands with the same "spring constant" together, you divide their combined spring constant by two. Therefore, we can use this system as a "divider."

- **Parallel Connections of Rubber Bands:**

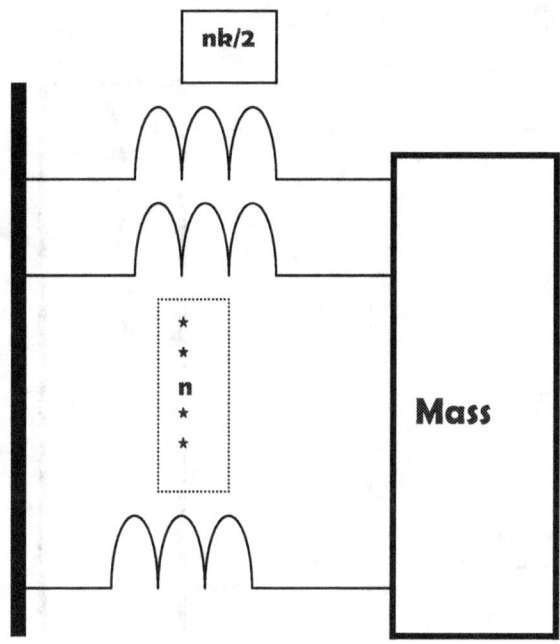

nk/2

Mass

```
┌──────────────────────────────────────────────────────┐
│                      Figure 2.                         │
│        Parallel Connection for "n" spring constants    │
│             K Tp/2  = ½(nk)                            │
│                    "Multiplier"                        │
└──────────────────────────────────────────────────────┘
```

The Parallel Connection of "n" springs has a combined spring constant, K_{Tp} given by the following computation.

1. $K_{Tp}/2 = k/2 + k/2 + * * * (n) * * * +k/2 = nk/2$

2. Therefore, this configuration of rubber bands produces a "multiplier" when two or more rubber bands with the same spring constant are put in parallel.

- **Analog Computer**

In order to estimate the value of π , we must create an "Analog Computer" to divide 22 by 7 using rubber bands and maybe some shower rings to hold them all.

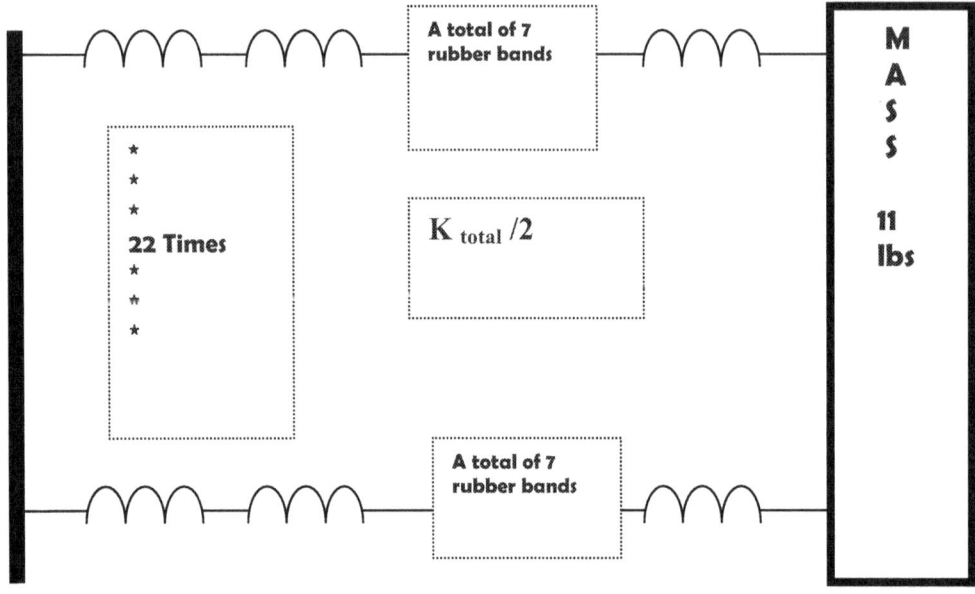

```
Figure 3.
The Analog Computer for π ~ 22/7
The displacement should be ~ 7 inches to make
the correct ratio.
```

The idea is to add 22 "divide by 7" circuits, so use 11 lbs of weight to cause an approximately 7 inch displacement. The ratio is then K/2 ~ 11/7 ~ $\pi/2$

- **Tensor Form of Hookes' Law:**

  F = -kX

$$\underline{F} = \begin{bmatrix} F1 \\ F2 \\ F3 \end{bmatrix} = - \begin{bmatrix} K11 & K12 & K13 \\ K12 & K22 & K23 \\ K13 & K23 & K33 \end{bmatrix} \begin{bmatrix} X1 \\ X2 \\ X3 \end{bmatrix}$$

Where 1,2,3 correspond to the coordinates x,y,z

Experiment #12: Acetone (Saying goodbye to Mr. Rubber Band!)

Just toss the rubber band into some acetone and see if you can find it? (It should dissociate!)
If you do this in a dark room, you might even see some "cosmic rays" in the acetone vapor.

Glossary of Terms:
Mechanics Terms:

1. **MASS:** A Quantity or aggregate of matter usually of considerable size.
   Symbolized by "M," units: Kilograms, Grams, or "Slugs"

2. **DISPLACEMENT:** The difference between the initial position of a body and a later position. Symbol usually ~ "x", units: meters, centimeters, or feet

3. **VELOCITY:** The time rate of motion in a given direction. Symbolized by "V"
   Units: Meters per second, centimeters per second, feet per second. (Our society has gotten used to a convenient unit for velocity called "Miles per Hour" which is not a standard unit is science, but is used in Engineering for practical convenience).

4. **ACCELERATION:** The act or process of changing velocity, the state of being accelerated. Symbol: A. Units: Velocity per unit time: meters per second squared, centimeters per second squared, or feet per second squared.

5. **FORCE:** The strength or energy exerted to cause a change in motion (F=MA).
   Units: Newtons, Dynes, and Pounds

6. **PRESSURE (Stress):** The Force applied across an area. (P = s = F/A). Units: Pascal (Newtons/square meter), (Dynes/square cm),
   Or Pounds per square inch (PSI)

7. **ELONGATION (Strain):** The state of being lengthened. (e=(x-x0)/ x0). Units: dimensionless

8. **MOMENTUM:** A property of a moving body that determines the length of time needed to bring it to rest. (p=MV). Units: Kilogram-meters per second, Gram centimeters per second, or Slug feet per second.

9. **WORK:** The sum of Force applied through a distance. (W=Fx).
   Units: Joules (Kg m/s2), Ergs (gm cm/ s2), or BTU (slug feet/ s2).

10. **KINECTIC ENERGY:** The energy associated with motion. (K=1/2mv2).
    Units: Joule, Ergs, and BTUs.

11. **POTENTIAL ENERGY:** The energy that apiece of matter has due to its position or the arrangement of its parts. (V=mgh)

12. **POWER:** 1.The ability to act or produce an effect. 2. The time rate of energy delivery. (P=dW/dt). Units: Watts (Joules per second), Ergs per second, or BTUs per sec.

13. **HARMONIC OSCILLATION:** 1. The act or fact of oscillating: Vibration; Fluctuation. 2. A flow changing periodically in direction. 3. A single swing (oscillation) from one extreme limit of motion to the other.

14. **MICROMETER CALIPER:** A caliper having a spindle moved by a finely threaded screw for making precise measurements.

15. **PERIOD OF OSCILLATION:** 1. The complete cycle of a series of events or a single action. 2. A portion of time determined by some recurring phenomenon.
    3. The interval of time required for cyclic motion to complete a cycle and begins to repeat itself. Symbol: T. Units: Seconds

16. **ANGULAR MOMENTUM:** 1. A vector quantity that is the measure of the intensity of rotational motion. 2. (L=Iw, where L is the angular momentum, I is the Moment of Inertia and w is the angular frequency). Units: Kg m2/sec, gm cm2/sec, or slug feet2/ sec.
    Handout: Physics Terms (Mechanics) Cont'd, P.2

17. **MOMENT OF INERTIA:** 1. The ratio of the torque applied to a rigid body produced about the axis of rotation to the angular acceleration thus produced about the axis. Symbol: I
    Units: Kilogram-meters, Gram-cm, or slug-feet

18. **TORQUE:** 1. To twist (as in torture!). 2. A Force, which produces or tends to produce rotation or torsion. (Torque= I (alpha), where torque is usually symbolized by the Greek letter "tau," I symbolizes the "moment of inertia" and the "angular acceleration" is symbolized by the Greek letter "alpha."
    Units: Kilogram meters2/ second2, Gram cm2/second2, Slug feet2/ seconds2.

19. **AMPLITUDE:** The extent of a vibratory motion measured from an equilibrium point to an extreme. Also see "displacement." Units: "length."

20. **NATURAL FREQUENCY:** The frequency that an object has when its oscillations are "unforced." (Frequency= 1/(2 times pi) times the square root of (k/m)).

21. HOOK's LAW: The law of "restoring forces" for conservative systems. ($F = -kx$, where F is the Force, k is the "spring constant" and "x" is the displacement.

22. STRESS STRAIN CURVE: A plotted graph showing the data collected from our experiment, which shows the "Energy" needed to break the rubber band as the area under the stress strain curve.

23. SPRING CONSTANT: A constant of proportionality that relates both force and displacement in Hook's Law.

24. NORMALIZATION: To make conform or reduce to a norm or standard (in the stress strain experiment we "normalize" our data so that we can compare the qualities and characteristics of the two different rubber bands equally).

25. SYMMETRY: 1. Balanced proportions. 2. The Beauty of form arising from balanced proportions. (The angular quantities are "symmetric with the "rectilinear" quantities).

26. DAMPING: The gradual reduction in amplitude of real oscillations. Symbol: (Greek letter: sigma).

27. WAVELENGTH: The distance in the line of advance of a wave from one point to another of constant phase. Symbol: (Greek letter: Lambda, Units: length(nm)).

28. FRACTURE: 1.The act or process of breaking or the state of being broken.

1. The rupture of soft tissue. 3. The result of fracturing. (Something happens in the stress strain test. What is it?).

29. CATASTROPHE: 1. The final event of the dramatic action, esp. of a tragedy. A momentous tragic event ranging from extreme misfortune to utter overthrow and ruin. (Why do you suppose that scientists developed "Catastrophe Theories?)?

30. POLYMER: A chemical compound or mixture of compounds formed by polymerization and consisting essentially of repeating structural units.

31. MICROSTRUCTURE: The microscopic structure of a material (as a mineral or biological cell).

Chemical Terms

1. "Polyisopropene" is the processed rubber derived from the naturally occurring butyl rubber that comes from rubber trees in the rainforests of South America. It has many useful properties including: elasticity and toughness that make it an ideal material for the mass production of tires for automobiles.
2. Rubber is also a "polymer," a kind long chain molecule made up of many repeated units. Polymers are also "elastomers," they are soft, yet sufficiently elastic that they rebound to their original shape.
3. Chemical processing is illustrative of one of the types of processing used to improve the properties of polymers. "Rubber Latex" is a suspension of polyisopropene obtained from certain plants that grow in tropical regions.
   "Hevea Rubber" –A naturally occurring rubber is composed of "Z-polyisopropene" and is a long hydrocarbon chain about 1,000 to 5,000 isopropene units in length. (See Fig.
4. The polyisopropene chains are randomly coiled and bound together by intermolecular "Van der Waals" forces ("Amorphic" or "Chaotic Structure")
5. Because the intermolecular forces are very weak, an external deforming force not only stretches the coiled polymer but also allows them to split past each other in a process known as "Plastic Flow."
6. When force is released, the polymer chains do not completely return to their original positions (A process called "Deformation").
7. In order to make natural rubber more elastic, it is heated with "sulfur" in a process known as "Vulcanization."
8. Mono-Sulfide and Di-Sulfide "Cross-links" form between the polyisopropene chains and provide sufficient "rigidity" to prevent plastic flow. Rubber usually has about 1-2% Sulfur. If too much cross-linking takes place, the rubber becomes a rigid solid and loses its elastomer properties.
9. TOUGHNESS: The energy needed to deform or break the material. It is also the area under the stress-strain curve.
10. PLASTIC STRAIN: The permanent displacement of atoms within the material that occurs upon exposure to higher stress.
11. ENERGY: of a gas of "point masses" is $E = 2/3 \, U$, U is the energy of an ideal gas ($U = 3/2kT$)

Appendix A: "Gnomon"
Introduction:

A "Gnomon" is a device that was used by the ancient Babylonians to measure the lengths of the year, by measuring the length of the sun's shadow each day, and on various special days, such as equinoxes and solstices when the alignment between the Sun and Earth made it possible to measure certain geometrical quantities like "latitude" and the "tilt of the earth" (latitude of the tropic).

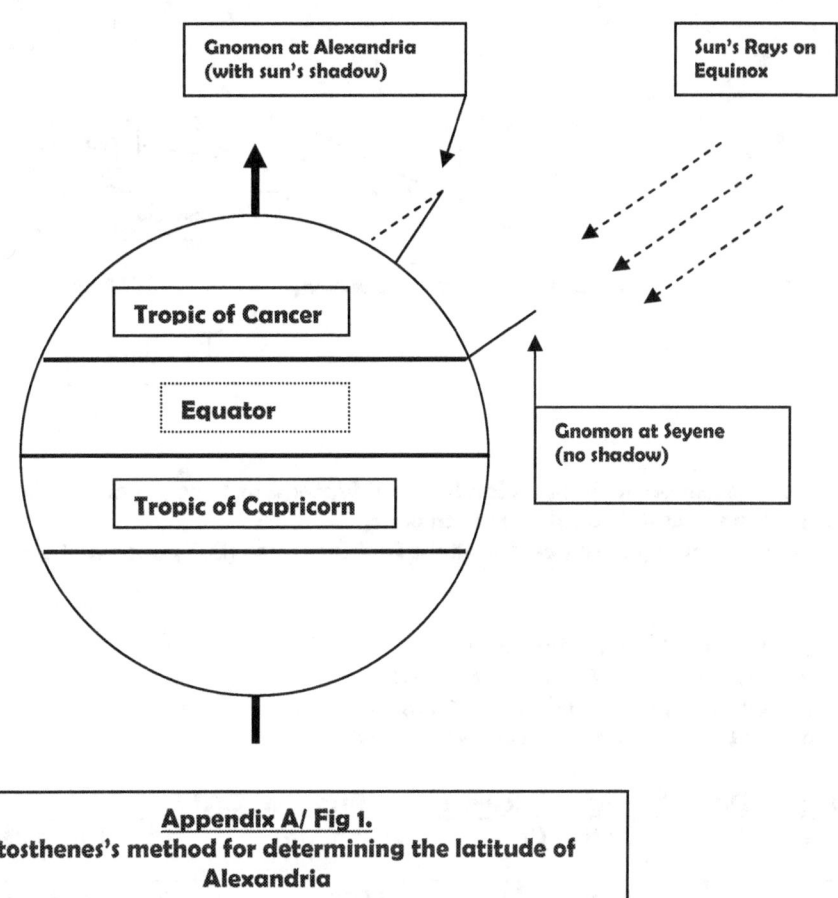

Gnomon at Alexandria (with sun's shadow)

Sun's Rays on Equinox

Tropic of Cancer

Equator

Tropic of Capricorn

Gnomon at Seyene (no shadow)

Appendix A/ Fig 1.
Eratosthenes's method for determining the latitude of Alexandria

Ancient geometers determined that the lengths of the congruent triangles formed on these days called the "equinoxes" revealed the latitude of the place, while the days called "solstices" (the longest and shortest days of the year) revealed the "tilt" of the earth relative to the sun.

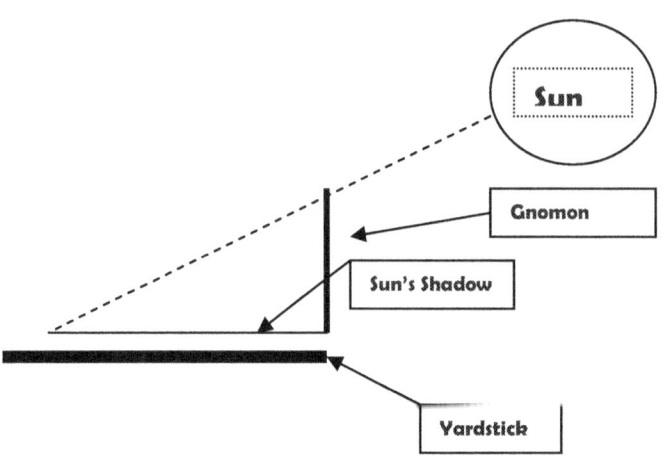

Sun

Gnomon

Sun's Shadow

Yardstick

Equinoxes occur on or around March 21 and September 21 each year, while Soltices occur on or around June 21 and December 21 each year.
Be sure to know if you are on "Daylight Savings Time" (DST) because there's a 1 hour shift!

Calculations:
1. Mean Latitude = Arctan (L/h)
2. Error = Arctan { (L +ΔL)/ (h - Δ h)}
3. Error Degrees = +/- (Error - Mean)
4. Latitude = Mean Latitude +/- Error degrees

Sample Data: Equinox 3-20-01 (Los Angeles, CA)

Time (PST)	Height (in)	Length (in)	Uncert +/- 0.25	Theta (+/- 0.8 deg)
11:45	34.0	23.500	0.014	34.6
11:50		23.375		34.5
11:55		23.250		34.2
12:00 (*)		23.125		34.1
12:05		23.000		34.1
12:10		23.000		34.1
12:15		23.000		34.1

Compare this value with the value for the latitude of Los Angeles: 34° 0' N

Collecting Data:

Time	Height	Length	Uncert	Theta

Questions:

1. What is the radius of the Earth using the value you have obtained for the latitude for your city? What else do you need to know?
2. how does this compare with the actual value?
3. Why was this such an important result in the ancient world?
4. What impact did this value have on the voyage of Christopher Columbus?
5. What direction is due South? Due North? Would you say that most buildings are solar aligned?
6. Now that you know the Earth's Radius, what are it's volume and surface area? If you assume that most of the Earth is made of Silicon with a density of 4.42 grams/cm^{-3}, what is the Earth's mass? How does this value compare with 6.0×10^{24} kg
7. How might a knowledge of your local latitude help you in building a house with passive "solar cooling?

Math note: (Universal Gravitation)

$F = GMm/r^2 = ma$ implies $\rightarrow$ $M = a\,r^2/G$

Where $a = 9.8$ m/sec^2, $r = 6.4 \times 10^6$ m, $G = 6.67 \times 10^{-11}$ m^3/ kg sec^2

This gives a result for the mass of the Earth of 6.0×10^{24} kg.

Appendix B: Math Physics Review: Simple and Angular Harmonic Oscillators

In physics, harmonic oscillators serve as models for how nature produces "vibrations." Using simple differential equations we can solve for all of the dynamical (i.e. motion related) quantities, such as displacement, velocity, acceleration, work, kinetic and potential energies, by solving a differential equation called the standard form of the "simple harmonic oscillator."

1. Force = F = ma (motion force)
2. Hooke's Law: F = -kx (rubber band restraining force, where k = spring constant)
3. Equating these two forces gives us; ma = -kx which can be rewritten into the standard form:
4. $m x'' + kx = 0$
5. Dividing by m gives $x'' + (k/m) x = 0$, where $\omega_o^2 = (k/m)$ so that
6. $x'' + \omega_o^2 x = 0$ (Is the standard form of the Harmonic Oscillator)
7. $\omega_o^2 = 2\pi / T = 2\pi f_o = k/m$
8. The natural frequency: $f_o = (1/2\pi) \sqrt{K/m}$
9. You choose the Amplitude "X_o"
10. The maximum velocity $V_{max} = 2\pi(X_o/ T)$
11. The maximum acceleration $A_{max} = (2\pi/T)^2 X_o$
12. From these quantities we can compute all of the others
    Max Force = $F_{max} = m A_{max} = (2\pi/T)^2 m X_o$
    Work = $W = F X_o = (2\pi/T)m X_o^2$
    Max Kinetic Energy = $K_{max} = \frac{1}{2} m V_{max}^2$
    Max Potential Energy = $U_{max} = mgX_o$
    $X(t) = X_o \cos \omega_o t$ or (exponential form) $X(t) = X_o \exp (-i \omega_o t)$
    $X'(t) = -\omega_o X_o \sin \omega_o t$ or $X'(t) = -i\omega_o X_o \exp (-i\omega_o t) = -i\omega_o X(t)$
    $X''(t) = -\omega_o^2 X_o \cos \omega_o t$ or $X''(t) = -\omega_o^2 X_o \exp (-i\omega_o t) = -\omega_o^2 X(t)$
13. All of these quantities coexist in the motion of the rubber band!

We can also show that this idea of simple harmonic oscillation works for the case of angular harmonic oscillations.

1. Torque = $\tau = - \Omega\theta$ (angular form of Hooke's law)
2. Torque = $\tau = I \theta''$ (Torque equals the moment of inertia, I and the angular acceleration θ'')
3. $I \theta'' + \Omega\theta = 0$
4. $\theta'' + (\Omega/I) \theta = 0$
5. $\theta'' + \omega_o^2 \theta = 0$ (the standard form of the angular harmonic oscillator: AHO)

6. Solutions:
    Angular displacement
    $\theta(t) = \theta_0 \cos \omega_0 t$ or (exponential form) $\theta(t) = \theta_0 \exp (-i\omega_0 t)$
    Angular velocity
    $\theta'(t) = -\omega_0 \sin \omega_0 t$ or (exp form) $\theta(t) = -i\omega_0\theta_0 \exp (-i\omega_0 t) = -i\omega_0 \theta(t)$
    Angular Acceleration
    $\theta''(t) = -\omega_0^2 \theta_0 \cos \omega_0 t = -\omega_0^2 \theta(t)$ or $\theta''(t) = -\omega_0^2 \theta_0 \exp (-i\omega_0 t) = -\omega_0^2 \theta(t)$
7. With these we can solve for everything
    Angular momentum: $L = I \theta' = -I \omega_o \sin \omega_0 t$
    Kinetic Energy = K $= \frac{1}{2} Ir^2 \theta'^2 = \frac{1}{2} I (r^2\omega_0 \sin \omega_0 t)^2$
    Torque $= \tau = I \theta'' = -I \omega_o^2 \theta_0 \cos \omega_0 t$
    Work = W $= \int_0^\infty \tau \, d\theta(t)$
    Potential Energy = U $= mg\theta = mg\theta_0 \cos \omega_o t$

The Laplace Transform solution for the Simple Harmonic Oscillator

1. $F = -bv - kx$ (equation of motion)

2. $mx'' + bx' + kx = 0$ (rewritten using derivatives)

3. $x(t) = A \exp(-ist)$ $\pounds \rightarrow X(s)$
   $x'(t) = s\, x(t)$ $\pounds \rightarrow -s\, X(s)$
   $x''(t) = s^2 x(t)$ $\pounds \rightarrow s^2 X(s)$ (taking the Laplace Transforms)

4. $(ms^2 - bs + k)\, X(s) = 0$ (plugging them back into equation 1.)

5. $s_{1,2} = \{-b +/- (b^2 - 4mk)^{1/2}\} / 2m$ (solving for the complex frequencies)

6. $x(t) = A \exp(is_1 t) + B \exp(-is_2 t)$ (the homogeneous solution)

Damping:

When you actually perform the experiments in this book, you will be confronted with the fact that in real experiments there will be "damping" of the motion, so I have decided that it would be appropriate to include a little bit about it. Let's start by writing the general form of the equation of motion.

1. $mx'' + bx' + kx = 0$ (general form of the equation of motion)
2. $x'' + (b/m) x' + (k/m) x = 0$
3. $s^2 + (b/m) s + k/m = 0$
4. $s_{1,2} = \{-(b/m) +/- \sqrt{(b/m)^2 - 4(k/m)}\}/ 2 = -(b/2m) +/- \sqrt{((b/2m)^2 - (k/m))}$

5. **Critical Damping Case:** (system goes back right away!)

   $(b/m)^2 = 4k/m$ or $(b/2m)^2 = k/m = \omega_0^2$ implying $\qquad m = b^2/4k$

6. **Over damping Case:** (system undergoes oscillations)

   $(b/m)^2 < 4k/m$ or $(b/2m)^2 < k/m = \omega_0^2$ which implies > $\quad m > b^2/4k$

7. **Under damping Case:** (System goes back only very slowly)

   $(b/m)2 > 4k/m$ or $(b/2m)2 > k/m = \omega_0^2$ which implies > $\quad m < b^2/4k$

Jordan Canonical Linear Form Solution for the Simple Harmonic Oscillator

1. $mx'' + bx' + kx = 0$ (standard form of the simple harmonic oscillator)
2. **Define new variables**

   $x = x_1$

   $x' = x_2 = x_1'$

   $x'' = x_3 = x_2'$

3. **Rewrite the differential equation**

   $x_2'' + (b/m) x' + kx = 0$

   or

   $x_2' + bx_1' + kx_1 = 0$

4. **Solve for the new derivatives:**
   - a) $\quad x_1' = -(k/b) x_1$
   - b) $\quad x_2' = -(k/b) x_1 - (b/m)x_2$

5. Plug the results from line 4 into the final form for the Jordan Canonical linear system (X' = AX +b (Linear matrix Form of the state equation))

6. $\begin{bmatrix} x'_1 \\ x'_2 \end{bmatrix} = \begin{bmatrix} -k & 0 \\ -k & -b \end{bmatrix}\begin{bmatrix} x_1 \\ x_2 \end{bmatrix} + \begin{bmatrix} b_1(0) = 0 \\ b_2(0) = 0 \end{bmatrix}$

   where $b_1(0) =$ intial velocity and $b_2(0) =$ initial acceleration.

7. $\Phi = \det (A - I\lambda) = 0$ (Characteristic equation)

   $= \det \begin{bmatrix} -(k+\lambda) & 0 \\ -k & -(b+-\lambda) \end{bmatrix}$

8. $\Phi = \lambda^2 + (b+k)\lambda - kb = 0$ (to solve for the eiggenvalues)

9. $\lambda_{1,2} = \{-(b+k) \pm \{(b+k)^2 + 4kb)^{1/2} \}/ 2$

10. $\Phi = \exp \begin{bmatrix} \lambda_1 \\ \lambda_2 \end{bmatrix}$ (matrix of eigenvalues)

11. $x(t) = \exp (At) = I + (1/1!) At + (1/2!) A^2 t^2 + ***$ (Solution by Cayley-Hamilton Theorem)

    $= 1 + \begin{bmatrix} -k & 0 \\ -k & -b \end{bmatrix}t + \frac{1}{2}\begin{bmatrix} k^2 & 0 \\ (k^2 +kb) & -b^2 \end{bmatrix} t^2 + ****$

12. $y(t) = \Phi x^T$

Coupled Oscillations:

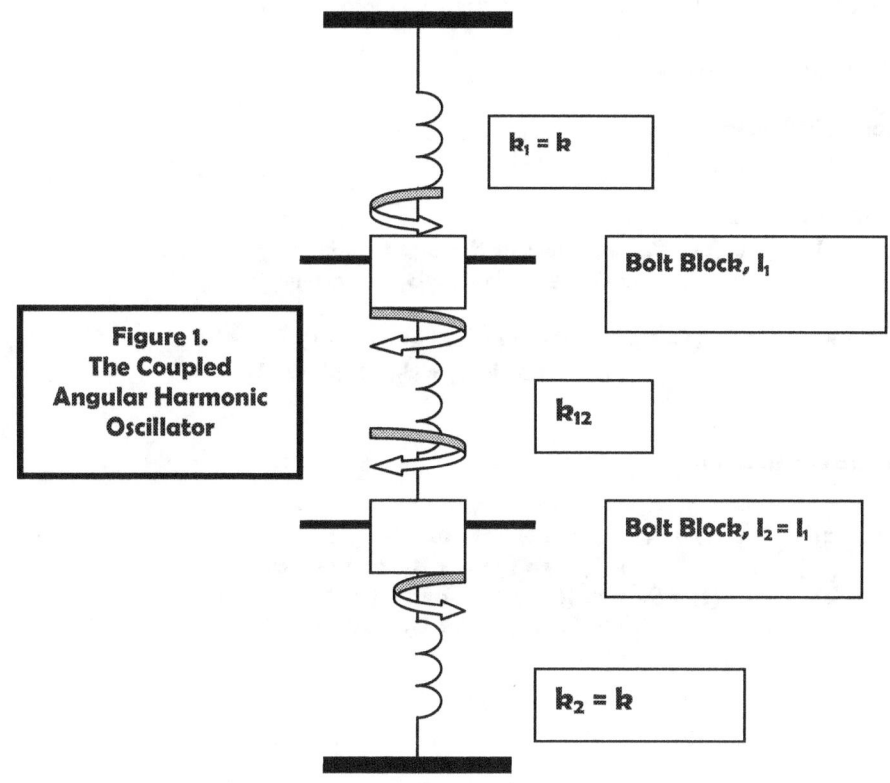

Figure 1.
The Coupled
Angular Harmonic
Oscillator

$k_1 = k$

Bolt Block, I_1

k_{12}

Bolt Block, $I_2 = I_1$

$k_2 = k$

---

1. **Equations of Motion**
   - c) $I\theta_1'' + (k + k_{12})\theta_1 - k_{12}\theta_2 = 0$
   - d) $I\theta_2'' + (k + k_{12})\theta_2 - k_{12}\theta_2 = 0$
2. **Solutions**
   - a) $\theta_1(t) = B_1 \exp(i\omega t)$
   - b) $\theta_2(t) = B_2 \exp(i\omega t)$

3. **Plug equation 2 into equation 1**
   - a) $-I\omega^2 B_1 \exp(i\omega t) + (k + k_{12}) B_1 \exp(i\omega t) - k_{12} B_2 \exp(i\omega t) = 0$
   - b) $-I\omega^2 B_1 \exp(i\omega t) + (k + k_{12}) B_2 \exp(i\omega t) - k_{12} B_1 \exp(i\omega t) = 0$

4. **Rewriting in terms of coefficients:**
   - a) $(k + k_{12} - I\omega^2) B_1 - k_{12} B_2 = 0$
   - b) $-k_{12} B_1 + (k + k_{12} - I\omega^2) B_2 = 0$

5. **Determinant of the coefficients B_1 & B_2:**

$$\text{Det} \begin{vmatrix} (k + k_{12} - I\omega^2) & -k_{12} \\ -k_{12} & (k + k_{12} - I\omega^2) \end{vmatrix} = 0$$

This leads to the "Secular Equation: $(k + k_{12} - I\omega^2)^2 - k_{12}^2 = 0$

88

6. **Hence**
   $(k + k12 - I\omega^2) = \pm k_{12}$
   Solving for ω (The Characteristic Frequency) yields:

   $\omega = \pm\sqrt{(k + k_{12} \pm k_{12})/I}$

   From which it follows that
   $\omega_1 = \sqrt{(k + 2k_{12})/I}$ **and** $\omega_2 = \sqrt{k/I}$

7. **General Solution**
   a) $\theta_1(t) = B_{11}^+ \exp(i\omega_1 t) + B_{11}^- \exp(-i\omega_1 t)$
         $+ B_{12}^+ \exp(i\omega_2 t) + B_{22}^- \exp(-i\omega_2 t)$

   b) $\theta_2(t) = B_{21}^+ \exp(i\omega_1 t) + B_{21}^- \exp(-i\omega_1 t)$
         $+ B_{22}^+ \exp(i\omega_2 t) + B_{22}^- \exp(-i\omega_2 t)$

8. **Particular Solution**

   a) $\theta_1(t) = B_1^+ \exp(i\omega_1 t) + B_1^- \exp(-i\omega_1 t)$
         $+ B_2^+ \exp(i\omega_2 t) + B_2^- \exp(-i\omega_2 t)$
   b) $\theta_2(t) = B_1^+ \exp(i\omega_1 t) + B_1^- \exp(-i\omega_1 t)$
         $+ B_2^+ \exp(i\omega_2 t) + B_2^- \exp(-i\omega_2 t)$

1. **Symmetric**

2. **Antisymmetric**

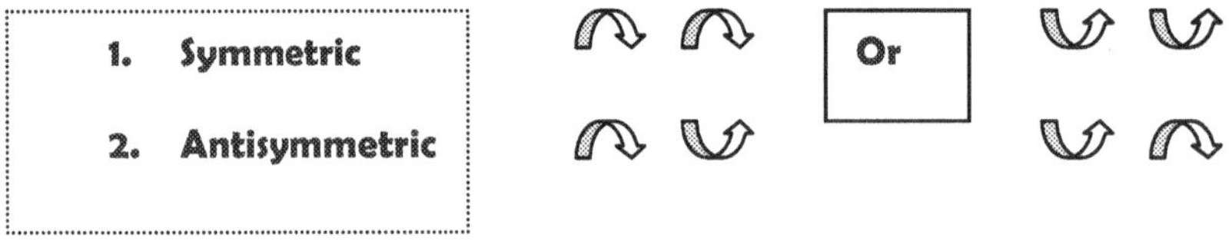

Figure 2.
Note that there are two modes: symmetric and antisymmetric

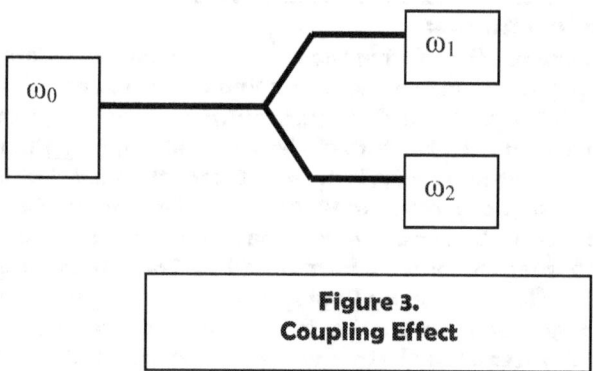

Figure 3.
Coupling Effect

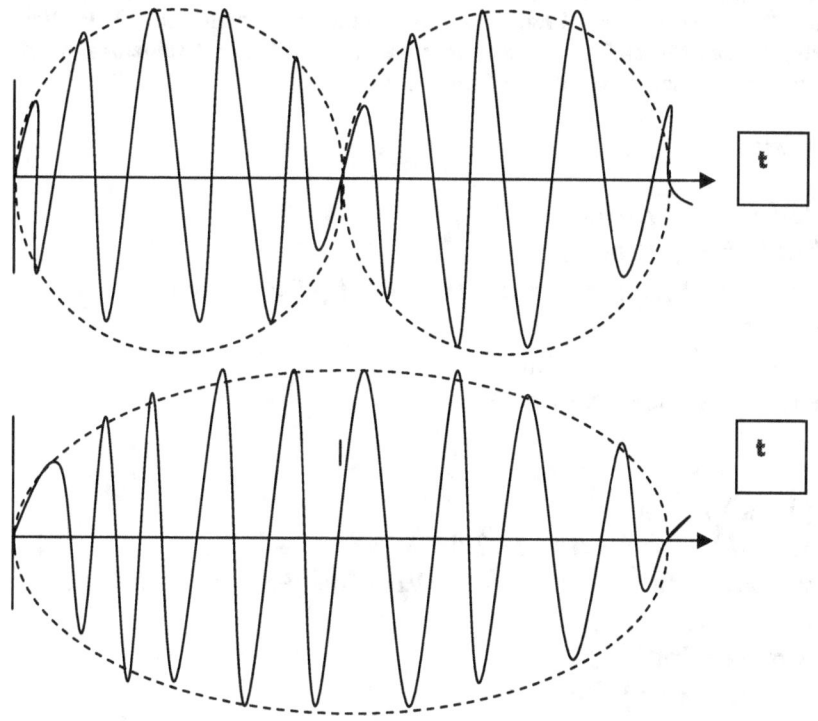

t

t

Figure 4.
"Beats" or Modulation for the two natural frequencies.

The Simple Harmonic Oscillator in Modern Physics :

A Short History of Quantum Mechanics:

Two rival theories competed with one another during the 19th century, one the "Corpuscular" theory of light, proposed by Newton, which viewed light as being composed of individual "particles," and the Thomas Young's "Wave theory" of light, which had been suggested from the results of certain interference experiments. At the turn of the 20th century, Max Plank and Albert Einstein proposed an even newer "Quantum theory," which put forth the idea that light has a dual nature, being both a "particle" and a "wave." In 1924, de Broglie suggested that all forms of matter exhibit the "dual behavior" of particles and waves. The mathematical description of the wave structure of atoms, and their interactions were formulated by Erwin Schrödinger, while the new "Quantum Mechanics" was later formulated by Paul A.M. Dirac and Werner Heisenberg. In this theory, the electron is no longer seen as a particle in orbit around a central nucleus, as was the case for the atom of Niels Bohr, instead the electron wave can extend all throughout the space of the atom, with the probability of its occurrence being highest were the Bohr radii were observed before. The quantization of these waves arises from this formulation, due to the discrete nature of the waves themselves, producing an integral number of waves around the atom through the probability density cloud of negative charge on a single electron. This allows the Schrödinger equation to be written for any atomic system, though exact solutions exist only for the first two elements: Hydrogen and Helium (Moore, p.450-1).

Creation Operators:

1. $\hat{a} = \beta/2 \, (x^\wedge + ip^\wedge/m\omega_0)$
2. $\hat{a}^+ = \beta/2 \, (x^\wedge - ip^\wedge/m\omega_0)$
3. $\hat{a} \neq \hat{a}^+$ is non-Hermitian as determined by $[x^\wedge, p^\wedge] = i\hbar$
4. $[\hat{a}, \hat{a}^+] = 1$
5. $\hat{a}\hat{a}^+ = 1 - \hat{a}^+\hat{a}$ (rewriting line 4)
6. $\hat{a}\hat{a}^+ + \hat{a}^+\hat{a} = 1$ (rewriting line 5)

Inverses:

7. $x^\wedge = (\hat{a} + \hat{a}^+) / \sqrt{2} \, \beta$
8. $p^\wedge = (m\omega_0/ i) \{ (a^\wedge - a^{\wedge+}) / \sqrt{2} \, \beta$
9. $\hat{H} = p^{\wedge 2}/ 2m + Kx^{\wedge 2}/2 = \hbar\omega_0 (\hat{a}^+\hat{a} + \tfrac{1}{2})$ **(The Hamiltonian for the Simple Harmonic Oscillator)**
10. $N^\wedge = \hat{a}\hat{a}^+$ **(definition)**
11. $H^\wedge \varphi_n = \hbar\omega_0 (N^\wedge + \tfrac{1}{2}) \varphi_n = \hbar\omega_0 (n + \tfrac{1}{2}) \varphi_n$
      (Taking the Hamiltonian of $\varphi_n = a_n \, Sin \, nkx$ or $a_n \, exp \, nkx$) (Particle in a Box Solution)

12. $< \varphi_n | H^\wedge \varphi_n > = E_n = \hbar\omega_0 (n + \tfrac{1}{2})$

13.   **Finally,**
      $E_n = \hbar\omega_0 (n + \tfrac{1}{2})$, n = 0, 1, 2, . . .
      (Remember we used this result in the section on phonons for the quantized energy levels)

Eigenfunctions of the Harmonic Oscillator Hamiltonian

1. $\hat{a} = \beta/2 \ (\ x^\wedge + ip^\wedge/m\omega_0 \ ) = \beta/2 \ (\ x^\wedge + \hbar/m\omega_0 \partial/\partial x \ ) = (2)^{-1/2} \ (\zeta + \partial/\partial\zeta \ )$

2. $\hat{a}^+ = \beta/2 \ (\ x^\wedge - ip^\wedge/m\omega_0 \ ) = \beta/2 \ (\ x^\wedge - \hbar/m\omega_0 \partial/\partial x \ ) = (2)^{-1/2} \ (\zeta - \partial/\partial\zeta \ )$

3. Where $\zeta^2 \equiv m\omega/\hbar \ x2 \equiv \beta^2 x^2$ is the non-dimensional displacement.

4. The time independent Schrödinger equation in one dimension may be expressed as

    $H^\wedge = p^{\wedge 2}/ \ 2m = \hbar^2/2m \ \partial^2/ \ \partial x^2 = E \ \varphi$, with $k^2 = 2mE/\hbar^2$

5. $\varphi_{xx} + k^2 \ \varphi = 0$ (Scalar time independent Schrödinger equation), where the subscript x denotes differentiation. Since there are no boundary conditions for "Free Particles" we arrive at the following solution

    $\varphi = A \ exp \ (ikx) + B \ exp(-ikx)$, which corresponds to energy eigenvalues given by

    $E = \hbar^2 k^2/ \ 2m = p^2/ \ 2m$

6. Rewriting equation 5 in terms of the non-displacement operator yields

    $(2 \ \hat{a}^+ \hat{a} + 1 - 2E/ \ \hbar\omega_0) \ \varphi = \varphi_{\zeta\zeta} + (\ 2E/ \ \hbar\omega_0 - \zeta^2) \ \varphi = 0$

7. The ground-state wave function φ_0 of the simple harmonic oscillator Hamiltonian obeys
    $a^\wedge\varphi_0 = 0$, which can be stated equivalently as

    $(\zeta + \partial/\partial\zeta)\varphi_0 = 0$, which in turn yields a solution: $\varphi_0 = A_0 \ exp \ (-\zeta^2/2)$, where it can be shown through normalization that $A_0 = \pi^{-1/4}$, then

8. $\varphi_0 (\zeta) = \pi^{-1/4} \ exp \ (-\zeta^2/2)$

9. The "n^{th}" eigenstate is given by

    $\varphi_n = A_n(\ \zeta + \partial/\partial\zeta)^n exp \ (-\zeta^2/2)$

10. And so the n th eigenstate of the simple harmonic oscillator Hamiltonian may be written:
    $\Phi_n = A_n \ \mathcal{H}_n (\zeta) \ exp \ (-\zeta^2/2)$ (Hamiltonian n^{th} eigenstate), where $\mathcal{H}_n (\zeta)$ are the "Hermite polynomials" generated by the recursion relation
    $A_n = (2^n \ n! \ \sqrt{\pi})^{-1/2}$ and are solutions to "Hermite's Equation":
    $\mathcal{H}_n (\zeta)'' -2\zeta \ \mathcal{H}_n (\zeta)' +2n \ \mathcal{H}_n (\zeta)$

11. The equation for the "n^{th} eigenvalue" may be written:

    $E_n = \hbar\omega_0 (n + 1/2)$

Natural Units:

The system of natural units is used in almost all quantum mechanics texts, where using energy units in Millions of Electron Volts (MeV), displacement in femtometers (fm), and time in seconds (sec), makes the problems easier to write and to visualize. Also, in this system we write $c = \hbar = 1$ to simplify the written expressions. With this in mind we can easily rewrite the final result for the quantized energy of the simple harmonic oscillator from the last section (equation 11.) in natural units as:

1. $E_n = \omega \, (n + \frac{1}{2})$ (MeV), where $n = \pm 1, \pm 2, \pm 3, \ldots$ ($\hbar = 1$)

As another example, we can rewrite the equation for the energy of a relativistic electron

2. $\quad E = [\ (pc^2) + (m_o c^2)\ ]^{\frac{1}{2}}$

More compactly in natural unit as:

3. $\quad E = [\ p^2 + m_o^2\ ]^{\frac{1}{2}}$ (MeV)

As a final example of the use of natural units, consider the formula for the radius of an atom:

4. $\quad r = A^{1/3}$ (fm), where A is the "Atomic weight" of the element atom and r is its radius.

X (Add- Step Barrier Potential; Liboff p. 256-65)
X (Add- WKB Approximation; Liboff p.232-5)

Bloch Wave Function

An electron wave propagating in a periodic potential (a lattice) may be described using a Bloch Wave Function, thus providing a set of states that can be shared by all the electrons in the lattice, but subject to Pauli exclusion.

The Schrödinger equation in one- dimension is (Cheo, p.120):

1. $\quad \partial^2\psi/\partial x^2 + 2m/\hbar\ [\ E - V(x)\ ]\ \psi = 0$
2. $\quad V(x) = \sum_{-\infty}^{+\infty} V_n \exp(-i2\pi nx/a)$, where a is the lattice spacing (amorphous).
3. $\quad \psi\,(x) = u(x) \exp(ikx)$ (Bloch Wave Function)
4. $\quad \psi\,(x) = B_o\,k_o \exp(ikx)$ (Free Particle)
5. $\quad E_n = \hbar\,k_o^2/\,2m$ (Energy of the Free Particle)
6. $\quad k_n = \pm\, n\pi/a$, where n = 0,1,2, ... (At the band edges)
7. $\quad \psi\,(x) = \exp(ikx)\,[\ B_o + \gamma B_n \exp(-i2\pi x/a)\ ] = B_o \exp(ikx) + \gamma B_n \exp(ik)\,k_n$
8. $\quad$ It can be shown that $k_o^2 = \frac{1}{2}\,[\ (k2 + kn2) \pm \sqrt{\{(k^2 - k_n^2) + 4\gamma^2 C_n^2\}}\ ]$
9. $\quad$ Then the Energy E(k) can be expressed
   $E(k) = \hbar^2/2m\,[\ k^2 + (\,k - 2\pi n/a\,)^2 \pm \sqrt{[\ k2 - (k - 2\pi n/a\,)^2\ ]^2 + (\,4m\,|Vn|\,/\,\hbar^2\,)^2}\ ]$ (parabola)

Appendix C: Proton Exchange Membranes and Fuel Cells

Proton Exchange Membranes (PEMs) form the core of a special kind of battery called a "Fuel Cell," which were once used almost exclusively for space missions, but have since undergone steady improvement until they are now ready to compete against even gasoline as a clean and cheap source of energy for everything from transportation to power generation.

The basic process taking place in a fuel cell involves dissociating H_2 gas into two H^+ ions, which then in turn migrate through the PEM to generate an electric current and then later, once on the other side of the PEM, creating pure water by recombining with Oxygen. This water can then serve as the fuel for the next cycle.

1. $2H_2O$ (l) $\rightarrow$ $2H_2$ (g)+ O_2 (g) (taking Hydrogen from water)
2. $2H_2$(g) $\rightarrow$ $4 H^+ + 4e^-$ (making electric current)
3. $4 H^+ + 4e^- + O_2$ (g) $\rightarrow$ $2H_2O$ (l)
   (recombining Hydrogen and Oxygen gases to make water again with no waste products).

The whole process is both vigorous and clean, a very good idea in a world in desperate need for new sources of energy. Most of the polymer science discussed in this book is also applicable to the study of Proton Exchange Membranes for Fuel Cells, which are identical to polymers like rubber except that they are "Fluorinated" instead of "Vulcanized." The whole design is a little like a dielectric capacitor except that in a fuel cell the Hydrogen gas most be fed into the manifold so that it can dissociate electrons and allow the protons to migrate (i.e. for the proton wave functions to be exchanged with the same quantum number (spin)).

4. $\Psi = \begin{bmatrix} 1 \\ 0 \end{bmatrix} \exp(ikx)$ or $\Psi = \begin{bmatrix} 0 \\ 1 \end{bmatrix} \exp(ikx)$

     (Proton (H^+) wave function with polarization (spin))

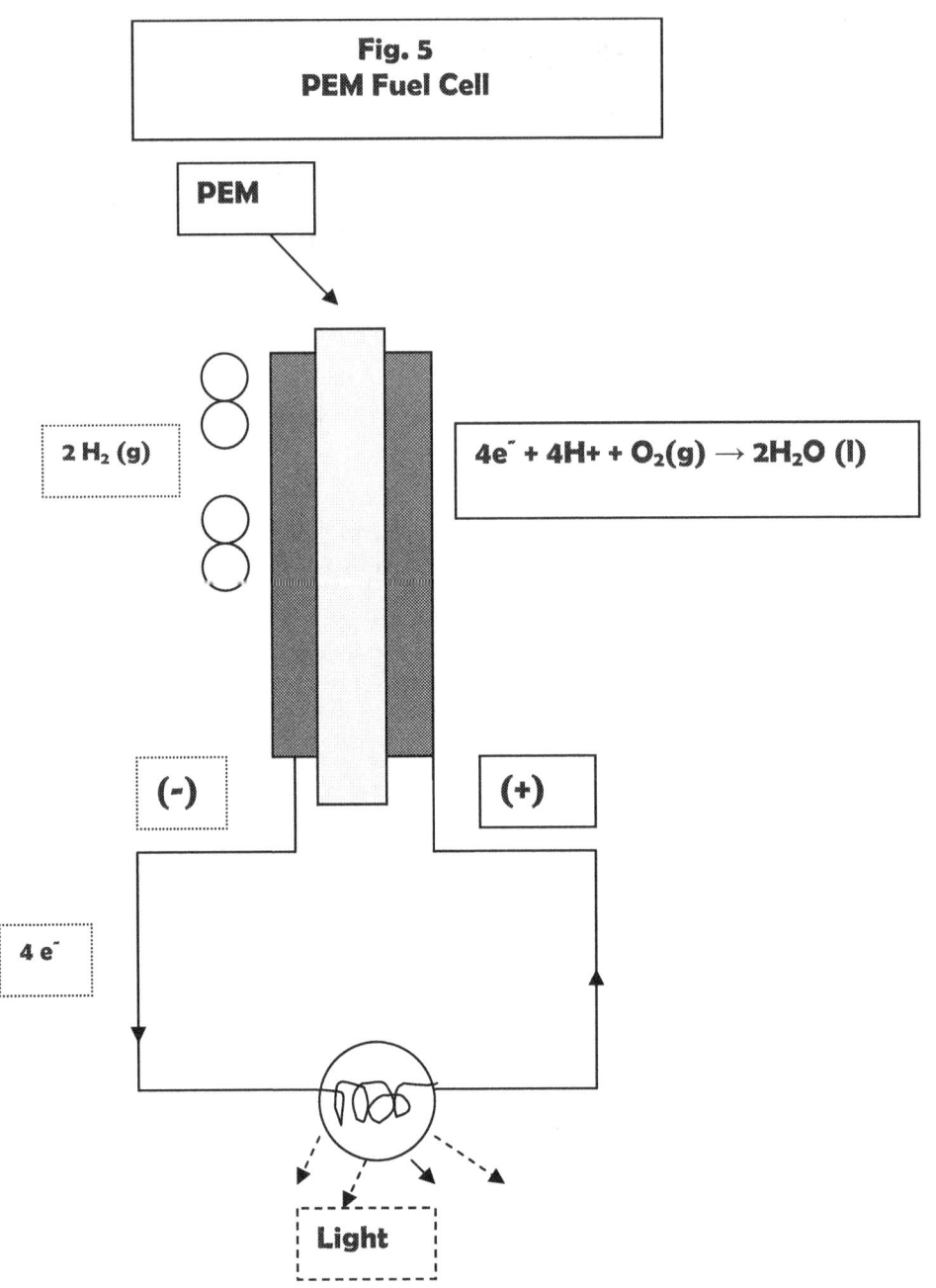

Fig. 5
PEM Fuel Cell

PEM

2 H₂ (g)

4e⁻ + 4H+ + O₂(g) → 2H₂O (l)

(-)

(+)

4 e⁻

Light

Appendix D: How to magnetize the nails for Experiment #9

The nails for the "Earth's Magnetic Field" experiment can be magnetized using the following set up: (use about 25 turns per nail)

Parts: 2 -6 Volt dry cell batteries
         4- finishing nails (~ 2 inches)
         5 feet of wire.

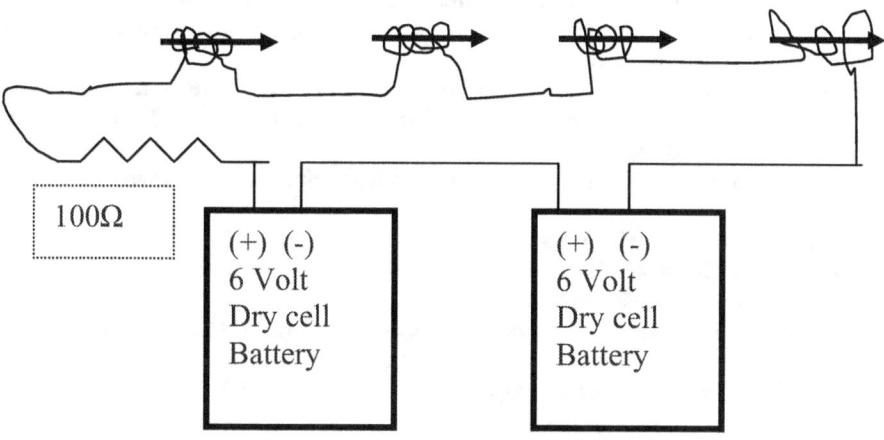

100Ω

(+) (-)
6 Volt
Dry cell
Battery

(+) (-)
6 Volt
Dry cell
Battery

Note :
Use 2 sets of 3 x 20 ohm, ½ Watt resistors, in parallel to make the load resistor.
This will give an equivalent resistance of 133 ohms.

Appendix E: How to Build Your Career

To become successful in working with science means doing all the same things you would in any other profession. Even with the loss of many high-tech defense jobs, there are still many companies seeking qualified applicants with both business and scientific training.

In addition to getting a good education, here are some ideas for how to build your own career.

Employers are looking for:
1. People who can organize, time, people, materials, budgets, and get tasks completed.
2. Effective team players, who will serve customers, and train other employees.
3. People who can quickly and accurately, assess information, analyze it and present it using computers, or other media.
4. People who bring a "can do" attitude towards the organization's success.
5. People who can make effective choices concerning large purchases, such as equipment, or other tools to accomplish assigned tasks.
6. People who know how to read critical information carefully and follow written instructions.
7. People who can speak and write effectively.
8. People who are good "listeners."
9. People who can make accurate computations
10. People who can make tough decisions based on good reasoning, think creatively, solve difficult problems, adapt to changes, and create useful solutions.
11. People who are "on-time" and show "respect for their fellow workers."

References:

Pre-Socratic Philosophers References:

1. Mc Kirahan, Richard D. (1994) "Philosophy before Socrates,"
   Hackett Publishing Company, Inc. (1994)
   Indianapolis, Indiana U.S.A.
   ISBN 0-87220-176-7
   B187.5.M35

2. By Plato, translated by Grube, G.M.A. and revised by Reeve, C.D.C (1992)
   "The Republic"
   Hacket Publishing Company, Inc. (1992)
   Indianapolis, Indiana U.S.A.
   ISBN 0-877220-137-6
   JC71.P35

3. Lindberg, David C. (1992) "Beginnings of Western Science"
   The University of Chicago Press (1992)
   Chicago, Illinois U.S.A.
   ISBN (paper) 0-226-48231-6
   Q124.95.L55

4. Popper, Karl; Paul, Routledge and Kegan (1963) "Back to the Presocratics,"
   from Conjectures and Refutations (paper)

5. Kirk, G.S. and Raven, J.E. (1957) "Thales of Miletus," extract
   from The Presocratic Philosophers
   Cambridge University Press (1957)

References Galileo and Newton:

1. Gingerich, Owen (1993) "The Eye of Heaven; Ptolemy, Copernicus, Kepler"
   The American Institute of Physics, New York

2. Ferguson, Kitty (2002) "Tycho & Kepler; The unlikely partnership that forever changed
   Our understanding of the heavens"
   Walker & Co, New York

3. Mizwa, P. & Stepen, P; A.M.., L.L.D. (1969) "Nicholas Copernicus (1543-1943)"
   Kennikat Press, Inc, Port Washington, New York

4. Chapin, Seymour L (1973) "Nicolaus Copernicus (1473-1973); His revolutions and His
   Revolution"
   Catalogue of an Exhibition of Manuscripts & Books with an historical Linderman Library,
   Lehigh University
   Bethlehem, Pennsylvania: September-November 1973

5. Adamczewski, Jan; "Nicolaus Copernicus and his Epoch"
   Written in cooperation with Piszek, Edward J.
   Copernicus Society of America
   Philadelphia, Pa. Washington D.C.
   Printed in Poland (A Corporate, Cultural, Public Program of
   "Mrs. Paul's" kitchens, Inc., Philadelphia, PA 19128

6. Sobel, Dava (1999) "Galileo's Daughter; A historical memoir of science, faith, and love"
   Walker & Co, New York

7. Goodman, David C. (1974) "The Conflict Thesis and Cosmology; Galileo and the Church"
(Written for the Course Team)
Open University Press (1974) (a distance learning package)

8. Szebehely, Victor G. (1989) "Adventures in Celestial Mechanics; A first course in the theory of orbits"
University of Texas Press, Austin
ISBN 0-292-75105-2
References Continued:
9. Westfall, Richard S. (1977) "The Construction of Modern Science; Mechanism and Mechanics"
The Cambridge History of Science Series
Cambridge University Press
ISBN 0-521-29295-6

10. Moore, Patrick; et al (1987) (editor) "The International Encyclopedia of Astronomy"
Orion Books, a division of Crown Publishers, Inc.
New York, N.Y.
ISBN 0-517-56179-4

11. Hetherington, Norris S. (1993) (editor) "The Encyclopedia of Cosmology; Historical, Philosophical, and Scientific Foundations of Modern Cosmology."
Garland Publishing, Inc.
New York, N.Y.
ISBN 0-8240-7213-8

References for Einstein:
1. Osserman, Robert (1999), "Poetry of the Universe"
All Rights Reserved.
References for Edwin Hubble:

1. Moore, P. (1987). "The International Encyclopedia of Astronomy,"
Orion Books, New York.

2. Osserman, R. (1995). "Poetry of the Universe,"
Doubleday, New York.

3. Zelik, M. & Smith, E (1987). "Introductory Astronomy & Astrophysics,"
Saunders College Publishing, New York.

4. Christianson, G. (1995). "Edwin Hubble: Mariner of nebulae,"
Farar, Strauss, and Giroux, New York.

References for Materials Science:
1. Tweeddale, J.G. (1973) "Materials Technology" Imperial College of Science & technology, London Published by Butterworth Group, London. ISBN 0-408-70391
2. Ashby, Michael F. & Jones, David H.R. "An Inbtroduction to microstructures, processing and design" Cambridge University, published by Pergammon Press International series on materials science & technology, volume 39.
3. Askeland, Donald (1994) "The Science and Engineering of Materials" Published by PWS Publishing Co. Boston.
4. Van Vlack, Lawerence H.(1980) "Elements of Materials Science and Engineering"

Addison-Wesley Publishing Co, Inc USA.
ISBN 0-201-08090-7

References for Mathematical Physics Review :

9. Liboff, Richard L. (1988), "Introduction to Quantum Mechanics"
    Addison-Wesley Publishers, 8th ed.
    ISBN 0-201-12221-9

    (For microstructure section)
10. Greiner,W; Neise, L.; Stocker, H (1995),
    "Thermodynamics and Statistical Mechanics"
    Springer-Verlag New York, Inc (all rights reserved)
    ISBN 3-540-94299 Springer-Verlag, N.Y.

11. Moore, Patrick; et al (1987) (editor) "The International Encyclopedia of Astronomy"
    Orion Books, a division of Crown Publishers, Inc.
    New York, N.Y.
    ISBN 0-517-56179-4

12. Marion, Jerry (1970) "Classical Dynamics- of particles and systems," 2nd ed.
    Academic Press
    Harcourt, Brace, Jovanovich, Publishers; New York, N.Y.
    Library of Congress catalog number
    78-107545
13. Cheo, P (1985) "Fiber Optics- Devices and Systems"
    Prentice-Hall, Inc., Englewood Cliffs, NJ 07632
    ISBN 0-13-314204-3

References for Catastrophe

14. Bowler, P. (2003) "Evolution- The history of an idea" (3rd ed.)
    University of California Press, Berkeley and Los Angeles
    ISBN 0-520-23693-9

Index:

www.ingramcontent.com/pod-product-compliance
Lightning Source LLC
Chambersburg PA
CBHW081138170526
45165CB00008B/2720